Elements of Mechanics

.n or before
low.

Elements of Mechanics

DAVID WILLIAMS

Loughborough University

Oxford   New York   Tokyo

OXFORD UNIVERSITY PRESS

1997

Oxford University Press, Great Clarendon Street, Oxford OX2 6DP

Oxford New York

Athens Auckland Bangkok Bogota Bombay Buenos Aires
Calcutta Cape Town Dar es Salaam Delhi Florence Hong Kong
Istanbul Karachi Kuala Lumpur Madras Madrid Melbourne
Mexico City Nairobi Paris Singapore Taipei Tokyo Toronto
and associated companies in
Berlin Ibadan

Oxford is a trade mark of Oxford University Press

Published in the United States
by Oxford University Press Inc., New York

© D. E. G. Williams, 1997

A catalogue record for this book is available from the British Library

Library of Congress Cataloging in Publication Data
(Data applied for)

ISBN 0 19 851881 1 (Hbk)
0 19 851880 3 (Pbk)

Typeset by Technical Typesetting Ireland
Printed in Great Britain by Bookcraft (Bath) Ltd., Midsomer Norton, Avon

Preface

The successful application of any science depends on the use of an appropriate mixture of principles and technique. Sometimes principles are easily understood, sometimes so difficult as to be opaque. In the case of simple mechanics the principles pose no particular mystery so that most students should feel quite happy that they have them readily to hand. Why then should there be an aversion to mechanics widespread amongst students who intend to become scientists or engineers? Maybe they know the principles but lack the technique required to apply those principles. The principles are simply useless lumber to be relegated to the attics of memory, remembered only with distaste. Often this rejection takes root in the dry mathematical form in which the subject is presented, unrelated to observation, quite irrelevant to life, and lacking any form of interest. No wonder the attitude, 'Shun the dessicated pedantry of mechanics', is met so frequently.

These days knowledge is presented in fifty minute quanta, or modules. There is no time available to drill students, to condition their responses to problems in mechanics. That was the old fashioned way which benefited some and discouraged many. A different approach is required, but what is it? Base the subject in observation rather than in apparently arbitrary rules. After all, mechanics, though heavily reliant on mathematics, developed as an observational science. Exposing the rationale underlying the 'laws' of mechanics serves also to show that the dogma of science, in the end, depends on observation. Setting problems based on a wide variety of experience should engage interest, challenge intelligence, and even stimulate curiosity. The fulfilment of one of these ambitions would be a success, of all of them would be a triumph.

Here then is my attempt to make mechanics palatable, to show its place as foundation of most of the science that students find more interesting. In making it I am indebted to all my predecessors in science but to no one of them in particular. It would be churlish not to tip my cap to them all, be they great men or small. I am grateful for the efforts of my editors at Oxford University Press, all of whom have been supportive and helpful. I tender my thanks to them.

Loughborough   D.W.
November 1996

To the reader

The mechanics with which we shall deal was developed as a science about three-and-a-half centuries ago. Whilst its formulation depended on the mathematical originality of Newton, its basis lay in the observations made by Galileo, Tycho, and many of their predecessors back to the days of the Greeks, or even the Egyptians and Babylonians. The success in races of a Formula 1 car, other things being equal, depends on the observations made by its driver being incorporated in mechanical adjustments to his vehicle. Mechanics isn't a treadmill mathematical exercise, it's an observational science. The examples, exercises, and problems that follow are couched in terms which relate them, even if only remotely, to everyday life.

It is my expectation that in following the text you will work your way through the examples and, at the end of the chapter, use that experience in attempting to solve the exercises. Some of them are interesting and some are tough. When you've done that, check your solutions against the ones given. Those solutions aren't, in many cases, the only ones available and may not be the best ones. It could well be that you will have produced a correct and more compact solution, providing a good reason for self-congratulation. The problems are given as additional exercises or for revision exercises; solve them after the chapter exercises or leave them for revision.

It is my hope that you will leave this exposition of mechanics with a clear view of the principles of mechanics and a grasp of the techniques for solving problems in mechanics. The realization will have dawned on you that mechanics underlies much of the rest of science and engineering. Then you can put this text on your bookshelf.

D.W.

Contents

Mathematical aide-memoire

1. Trigonometry

$\sin \theta = -\sin(-\theta) = \sin(\pi - \theta) = \cos(\frac{1}{2}\pi - \theta)$ $(1/\sin \theta) = \mathrm{cosec}\, \theta$

$\cos \theta = \cos(-\theta) = -\cos(\pi - \theta) = \sin(\frac{1}{2}\pi - \theta)$ $(1/\cos \theta) = \sec \theta$

$\tan \theta = (\sin \theta/\cos \theta) = -\tan(-\theta)$

$(1/\tan \theta) = \cot \theta$

$\tan \theta = \cot(\frac{1}{2}\pi - \theta)$

$\sin^2 \theta + \cos^2 \theta = 1$

$1 + \cot^2 \theta = \mathrm{cosec}^2\, \theta$

$\tan^2 \theta + 1 = \sec^2 \theta$

$\sin(\theta + \phi) = \sin \theta \cos \phi + \cos \theta \sin \phi$

$\cos(\theta + \phi) = \cos \theta \cos \phi - \sin \theta \sin \phi$

$\tan(\theta + \phi) = \dfrac{\tan \theta + \tan \phi}{1 - \tan \theta \tan \phi}$

$\sin^2 \frac{1}{2}\theta = \frac{1}{2}(1 - \cos \theta)$ $\cos^2 \frac{1}{2}\theta = \frac{1}{2}(1 + \cos \theta)$

$\tan^2 \frac{1}{2}\theta = \dfrac{1 - \cos \theta}{1 + \cos \theta}$

$\sin \theta + \sin \phi = 2 \sin[\frac{1}{2}(\theta + \phi)]\cos[\frac{1}{2}(\theta - \phi)]$

$\sin \theta - \sin \phi = 2 \cos[\frac{1}{2}(\theta + \phi)]\sin[\frac{1}{2}(\theta - \phi)]$

$\cos \theta + \cos \phi = 2 \cos[\frac{1}{2}(\theta + \phi)]\cos[\frac{1}{2}(\theta - \phi)]$

$\cos \theta - \cos \phi = 2 \sin[\frac{1}{2}(\theta + \phi)]\sin[\frac{1}{2}(\phi - \theta)]$

2. Triangles

For the triangle shown in Fig. M1,

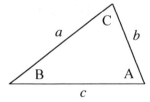

Fig. M1 The triangle with sides a, b, and c, and angles A, B, and C.

the sum of the angles $A + B + C = 180° = \pi$ radians.
The cosine rule states that

$$a^2 = b^2 + c^2 - 2bc \cos A$$
$$b^2 = a^2 + c^2 - 2ac \cos B$$
$$c^2 = a^2 + b^2 - 2ab \cos C$$

and the sine rule that

$$\frac{a}{\sin A} = \frac{b}{\sin B} = \frac{c}{\sin C}.$$

3. Differentiation

Function, $f(x)$	Derivative, $(\mathrm{d}[f(x)]/\mathrm{d}x)$
a constant, a	0
ax	a
ax^n	anx^{n-1}
(where n is a real number which may be positive or negative)	
$\sin ax$	$a \cos ax$
$\cos ax$	$-a \sin ax$
$\tan ax$	$a \sec^2 ax$
$\cot ax$	$-a \operatorname{cosec}^2 ax$
$\sec ax$	$a \tan ax \sec ax$
$\operatorname{cosec} ax$	$-a \cot ax \operatorname{cosec} ax$
$e^{ax} \; (= \exp(ax))$	$a \exp(ax)$
$\ln ax \; (= \log_e ax)$	a/x

If the function is a product of two other functions, i.e. $f(x) = g(x)h(x)$ then

$$(\mathrm{d}[f(x)]/\mathrm{d}x) = (\mathrm{d}f/\mathrm{d}x) = g(\mathrm{d}h/\mathrm{d}x) + h(\mathrm{d}g/\mathrm{d}x).$$

A derivative may be expressed in terms of a derivative with respect to another variable as a product, e.g.

$$\frac{\mathrm{d}f}{\mathrm{d}x} = \frac{\mathrm{d}f}{\mathrm{d}y} \frac{\mathrm{d}y}{\mathrm{d}x}.$$

4. Integration

Function	**Indefinite integral** (constant to be added)

$f(x)$ $\qquad\qquad\qquad\qquad\qquad\qquad\qquad\qquad \int f(x)\,dx$

x^n $\qquad\qquad\qquad\qquad\qquad\qquad\qquad [1/(n+1)]x^{n+1}$

$(1/x)$ $\qquad\qquad\qquad\qquad\qquad\qquad\qquad\qquad \ln(x)$

$\exp(ax)$ $\qquad\qquad\qquad\qquad\qquad\qquad\qquad (1/a)\exp(ax)$

$\sin(ax)$ $\qquad\qquad\qquad\qquad\qquad\qquad\quad -(1/a)\cos(ax)$

$\cos(ax)$ $\qquad\qquad\qquad\qquad\qquad\qquad\quad (1/a)\sin(ax)$

$\tan(ax)$ $\qquad\qquad\qquad\qquad\qquad\qquad -(1/a)\ln[\cos(ax)]$

$\cot(ax)$ $\qquad\qquad\qquad\qquad\qquad\qquad (1/a)\ln[\sin(ax)]$

$\sec(ax)$ $\qquad\qquad\qquad\qquad (1/2a)\ln[\sec(ax)+\tan(ax)]$

$\dfrac{1}{x^2+a^2}$ $\qquad\qquad\qquad\qquad\qquad (1/a)\tan^{-1}(x/a)$

$\left(\dfrac{1}{a^2+x^2}\right)^{1/2}$ $\qquad\qquad (1/a)\sinh^{-1}(x/a)\qquad$ for $(a^2-x^2)>0$

$\dfrac{1}{x^2-a^2}=(1/2a)\left(\dfrac{1}{x-a}-\dfrac{1}{x+a}\right)$ $\qquad (1/2a)\ln\left(\dfrac{x-a}{x+a}\right)$

$\cos^n(x)\sin(x)$ $\qquad\qquad\qquad\qquad -[1/(n+1)]\cos^{n-1}(x)$

$\qquad\qquad\qquad$ because $[-\sin(x)]$ is the derivative of $\cos(x)$

5. Vectors

Vectors are generally signified by bold type, e.g. a, and their magnitude by normal type, e.g. a. In terms of the components of a along the x-, y-, and z-axes

$$a = a_x\mathbf{i} + a_y\mathbf{j} + a_z\mathbf{k}$$

with $\mathbf{i}$, $\mathbf{j}$, and $\mathbf{k}$ the unit vectors in the x-, y-, and z-directions.

The scalar product of two vectors a and b, $a\cdot b$, is a scalar quantity of magnitude $ab\cos\theta$, where θ is the angle between the directions of a and b.

As an example, the scalar products of the unit vectors have the forms $\mathbf{i}\cdot\mathbf{i}=1$ and $\mathbf{i}\cdot\mathbf{j}=\mathbf{i}\cdot\mathbf{k}=0$.

The vector product of two vectors a and b, $a\times b$ $(=c)$, is a vector of magnitude $ab\sin\theta$, where θ is the angle between the directions of a and b. c is directed perpendicular to the plane which contains a and b. Its direction with respect to that plane is defined by the direction in which a right-hand threaded screw (a corkscrew) would advance if it were rotated from a to b. This definition leads to the identity

$$a\times b = -b\times a \qquad \text{or} \qquad a\times b + b\times a = 0.$$

The components of a vector, $c = a \times b$, have the form

$$c_x = (a_y b_z - a_z b_y)$$
$$c_y = (a_z b_x - a_x b_z)$$
$$c_z = (a_x b_y - a_y b_x)$$

where the a_x's, etc., are the components of a and b.

6.   Summary of mechanics formulae

Vectors

$$r = |r| = |xi + yj| = (x^2 + y^2)^{1/2}$$
$$R = r_1 + r_2 = (x_1 + x_2)i + (y_1 + y_2)j$$
$$|R| = |r_1 + r_2| = \left[(x_1 + x_2)^2 + (y_1 + y_2)^2\right]^{1/2}$$
$$R^2 = r_1^2 + r_2^2 - 2r_1 r_2 \cos(180 - \alpha) = r_1^2 + r_2^2 + 2r_1 r_2 \cos \alpha$$
$$r_1 \cdot r_2 = r_1 r_2 \cos \alpha$$
$$R^2 = |R|^2 = R \cdot R = r_1^2 + r_2^2 + 2r_1 \cdot r_2$$
$$= |r_1|^2 + |r_2|^2 + 2|r_1||r_2|\cos \alpha$$
$$\frac{R}{\sin \alpha} = \frac{r_1}{\sin \beta} = \frac{r_2}{\sin \gamma}$$

Linear motion

$$v = iv_x + jv_y = i\frac{dx}{dt} + j\frac{dy}{dt}$$
$$x_2 = Vt + x_1$$
$$a_x = \frac{dv_x}{dt}; \quad a_y = \frac{dv_y}{dt}; \quad a_z = \frac{dv_z}{dt}$$
$$a = |a| = \left(a_x^2 + a_y^2\right)^{1/2}$$
$$a = v^2/R$$
$$a = \frac{dv}{dt} = \frac{d(dx/dt)}{dt} = \frac{d^2x}{dt^2}$$
$$a = \frac{dv}{dt} = \frac{dv}{dx}\frac{dx}{dt} = v\frac{dv}{dx}$$
$$v_2 = v_1 + at$$
$$x = v_1 t + \tfrac{1}{2}at^2$$
$$v_2^2 = v_1^2 + 2ax$$

Centre of mass

$$m_1 x_1 + m_2 x_2 = (m_1 + m_2)X$$

$$X = \frac{\sum m_i x_i}{\sum m_i}$$

$$V = \frac{m_1 v_1 + m_2 v_2}{m_1 + m_2}$$

Momentum, force, etc.

$$p_1 + p_2 = m_1 v_1 + m_2 v_2 = \text{constant}$$

In the centre of mass frame of reference:

$$p_{g1} + p_{g2} = 0$$

$$F_1 = \frac{dp_1(t)}{dt}$$

$$F_1 = m_1 \frac{dv_1}{dt} = m_1 v_1 \frac{dv_1}{dx_1} = m_1 \frac{d^2 x_1}{dt^2}$$

$$F_1 \, dt = dp_1$$

$$\int F_1 \, dt = p_1(t) - p_1(0)$$

$$J_2 = p_2 = -p_1 = -J_1$$

$$F_1 = -F_2$$

$$|F_1 x| = \tfrac{1}{2} m_1 v_1^2(x)$$

$$F \cdot ds = F_x \, dx + F_y \, dy + F_z \, dz$$

$$= \tfrac{1}{2}\left[mv_x^2 + mv_y^2 + mv_z^2 \right] = T$$

$$T = p^2/2m$$

$$P = \frac{d}{dt}(F \cdot r) = \frac{dW}{dt}$$

$$F \cdot (dr/dt) = F \cdot v$$

$$W = \int P(t) \, dt$$

$$F = pA$$

$$a \times b = i(a_y b_z - a_z b_y) + j(a_z b_x - a_x b_z) + k(a_x b_y - a_y b_x)$$

$$H = T + V$$

Conservative force

$$F_y = -\frac{d}{dy}[V(y)]$$

Projectiles

$$y = x \tan \theta - \frac{gx^2}{2v^2 \cos^2 \theta}$$

SHM

$$v = \omega(a^2 - x^2)^{1/2}$$
$$\sin^{-1}(x/a) = \omega t + C'$$
$$\mathcal{T} = 2\pi/\omega$$
$$\tfrac{1}{2}m\omega^2 a^2 = \tfrac{1}{2}mv^2 + \tfrac{1}{2}m\omega^2 x^2$$

Rotational motion

$$I = \lim_{\delta m \to 0} \sum r^2 \, \delta m = \int r^2 \, dm$$
$$\Gamma = I(d\omega/dt) = I\omega(d\omega/d\theta) = I(d^2\theta/dt^2)$$
$$L = I\omega$$
$$\Gamma = (dL/dt)$$
$$J = \int \Gamma \, dt = \delta L$$
$$W = \Gamma\theta$$
$$P = \Gamma\omega$$
$$I_A = I_C + Ma^2$$
$$I_Z = I_Y + I_X$$
$$\Gamma = r \times F$$
$$v = \omega \times r$$
$$l = m(r \times \omega \times r) = r \times mv = r \times p$$
$$\frac{dv_r}{dt} = \frac{d^2 r}{dt^2} - r\left(\frac{d\theta}{dt}\right)^2$$
$$\frac{dv_\theta}{dt} = \frac{1}{r}\frac{d}{dt}\left(r^2 \frac{d\theta}{dt}\right)$$
$$\Gamma = \frac{dL}{dt}$$
$$T = \tfrac{1}{2}\left(I_x \omega_x^2 + I_y \omega_y^2 + I_z \omega_z^2\right)$$

A preamble on dimensions and units

Dimensional homogeneity of equations

The practice of science could be described as the process of observation followed by the construction of verbal or mathematical models to explain the observations, or vice versa. The use of mathematical models implies that the observations are quantitative, that they involve numbers used to specify the observations. Simple observations—for example, about mortality statistics—involve counts of the number of deaths and of the total population, and those numbers are simply numbers providing an answer to the question, 'How many are there?'. However, simple numbers soon become inadequate. Asking how far it is from El Paso to Houston and receiving the reply 'It's about 750', just leads to confusion. The answer '750' means 750 miles in American/Imperial measure, but unless 'miles' is specified the number could be a count of any other measure of length—furlongs, kilometres, leagues, or even ells. But the Scottish, the English, and the Irish ells are all of different lengths. Equally, the answer might have been given in minutes travelled in a helicopter, or maybe in gallons of fuel used by a particular make and model of car. But not all helicopters travel at the same speed, and the US gallon is 17% less than the Imperial gallon . . . .

How far is it from St. Louis to Amarillo? About 750, but this time, because we've realized the ambiguities involved in using only the number, we attach it to a measure of length to produce the answer 'It's 750 miles'. We've associated a unit of length with the number to make the statement of distance comprehensible to anyone familiar with the Imperial/US standard of length. The distance 750 has been measured in units of length—miles— and we say that the distance 750 miles has the dimension of length. This means that we can compare the distance only with another distance or length. If we have an equation which has distance on its left-hand side, then the dimension of its left-hand side is length and its right-hand side must have the dimension of length as well. As an illustration of this statement, let's introduce the metre (m) as unit of length and the kilometre (1 km = 1000 m) as its big brother. This takes us within the conventional scientific system of units, the Système International (SI), in which the metre is the unit of length. We know that the distances measured above were in miles, and if we want to communicate with someone who knows only SI units we have to convert miles to metres, or more sensibly to kilometres. The number of miles in a kilometre is about 0.625 and so we write

$$750 \text{ miles} = (750/0.625) \text{ km} = 1200 \text{ km}.$$

One length has been identified with another. The equation is described as dimensionally homogeneous. If we denote the dimension 'length' by [L] we can express this homogeneity as [L] = [L], seemingly a trivial statement but nonetheless significant. The number of miles per km is expressed dimensionally by the ratio [L]/[L]. The [L]'s top and bottom cancel out so that the number of miles per km is a number without dimensions, usually called a number but sometimes, more specifically, a dimensionless number.

Let's look again at the distance, expressed unconventionally, as above, in terms of gallons of fuel by a car. The gallon is a measure of volume, which certainly doesn't have the dimension [L], so that the equation relating gallons to miles has to be more complicated dimensionally than the equation relating km to miles. To convert gallons to miles in this case we have to multiply the number of gallons by the fuel consumption of the vehicle. This consumption is measured in miles travelled on a gallon of fuel, and the equation that we use has the form

$$\text{miles} = \text{gallons} \times (\text{miles per gallon}) = \text{gallons} \times \frac{\text{miles}}{\text{gallons}}.$$

The quantity 'gallons' cancels top and bottom in this equation, leaving only miles, just as is required by the necessity that equations are dimensionally homogeneous. We can go further than this by making use of the observation that a gallon (like a cubic metre, a unit volume expressed in SI units) is a volume. We know that volume–for example, for a rectangular parallelopiped—may be obtained by multiplying its length by its breadth to give the area of its base, and then multiplying that area by its depth. Two points arise from this observation; area is obtained by multiplying together two lengths, and volume by multiplying area by length. We can express these statements dimensionally as

$$[\text{Area}] = [L][L] = [L]^2, \qquad [\text{Volume}] = [L][L]^2 = [L]^3,$$

where the square brackets around 'Area' and 'Volume' indicate that we are defining their dimensions. The equation converting gallons to miles can be written dimensionally as

$$[L] = [L]^3 \left\{ [L]/[L]^3 \right\} = [L]$$

which is, of course, dimensionally homogeneous.

The other fundamental quantities of dimensional interest are time, [T], and mass [M]. The SI units of time and mass are seconds and kilograms, respectively. It's easy enough to derive the dimensions of the quantities used in mechanics; for example, velocity is measured in metres per second, so that its dimensions are

$$[\text{Velocity}] = [L][T]^{-1}.$$

Velocity appears first in Chapter 2. This statement leads on to a table of dimensions of quantities occurring in mechanics (the third column of the table is a reference to the chapter in which the quantity is used):

Quantity	Dimensions	Chapter
[Acceleration]	[L] [T]$^{-2}$	3
[Momentum]	[M] [L] [T]$^{-1}$	5
[Force]	[M] [L] [T]$^{-2}$	5
[Energy]	[M] [L]^2 [T]$^{-2}$	5
[Torque]	[M] [L]^2 [T]$^{-2}$	10
[Angle]	(—dimensionless)	
[Angular velocity]	[T]$^{-1}$	3
[Angular momentum]	[M] [L]^2 [T]$^{-2}$	10
[Angular acceleration]	[T]$^{-2}$	10
[Moment of inertia]	[M] [L]^2	10

Dimensional analysis

Example. It seems plausible that the period of oscillation of a pendulum should depend on its mass. By using dimensional arguments, show that the period of oscillation is in fact independent of the pendulum mass.

The pendulum has mass and length. We may write a dimensional equation relating the period of oscillation to these quantities as

$$[\text{Period}] = [\text{Mass}]^x\,[\text{Acceleration}]^y\,[\text{Length}]^z,$$

where x, y, and z are unknown exponents. The equation can be written as

$$[T] = [M]^x[LT^{-2}]^y[L]^z = [M]^x[L]^{(y+z)}[T]^{-2y}.$$

If the equation is to be dimensionally homogeneous, as it must be, then

$$x = 0, \qquad y + z = 0, \qquad -2y = 1,$$

which leads to

$$x = 0, \qquad y = -\tfrac{1}{2}, \qquad z = \tfrac{1}{2}.$$

This means that mass does not appear in the expression for the period of oscillation, and thus,

$$\text{Period} \propto \left(\frac{\text{Length}}{\text{Acceleration due to gravity}}\right)^{1/2}.$$

This dimensional analysis leaves us completely in the dark about the nature of the constant which would make the proportional sign, α, into an equality. All we can say is that the constant must be dimensionless. An example of a dimensionless constant made up from quantities which have dimensions is the quantity ωt, which occurs in studies of the linear oscillator: ωt is the product of an angular velocity ω, the dimension of which is [T]$^{-1}$, and time, t, the dimension of which is [T]. The product of these has dimensions [T]^0 ($= [T][T]^{-1}$), i.e. it is a number, without dimensions.

How do we deal with functions such as sines and exponentials? The first definition of sine that we're likely to come across is 'the ratio of the lengths of the side opposite to the angle and of the hypotenuse', referring to a right angled triangle. The ratio of these lengths is a dimensionless quantity. The angle, θ, is a dimensionless quantity, and this leads us to observe that the argument, x, of any function such as $\sin x$ must be a dimensionless quantity. If the argument involves quantities which have dimensions then those dimensions must cancel out to make the argument of the function dimensionless. The product ωt, with ω as angular velocity (or angular frequency) and t as time, is one such dimensionless quantity, i.e. the quantity $\sin \omega t$ is dimensionally correct. Again, for example, in the exponential function $\exp(-kv^2)$, with v a velocity, the quantity k must have dimensions which cancel out those of v^2, i.e. $[k] = [L^2 T^{-2}]^{-1}$, while in the exponential function $\exp(-Kv^2/v_0^2)$, with v_0 a constant velocity, K is dimensionless.

Units

The units of time, length, and mass we use are those of the Système International, the second, s, the metre, m, and the kilogram, kg. The second is defined in terms of the frequency of a particular spectral line emitted by Cs^{135} atoms. The metre is then defined in terms of the distance travelled by light in vacuum in a specified time. These two standards are properties of matter and electromagnetic radiation, independent of any man-made standards. They're very reproducible and, as far as can be established, stable. The unit of mass is defined as the mass of a billet of platinum/iridium alloy stored in very clean conditions in Paris. It's reproducible to the precision with which the purity of the alloy components, the composition of the alloy, and the shape and size of the billet can be reproduced by precise preparation and machining, but this standard unit is artificial and arbitrary rather than natural.

It's not always convenient to describe observations in terms of metres, kilograms, and seconds. Some events occur in time intervals which are almost instantaneous; some lengths are small and some are large, compared with a metre. In principle, this is allowed for by defining 'sub-units' as, for example, a thousand times larger than, or smaller than, the fundamental unit. The prefixes used to describe multiplication or division by these factors of 1000 are listed below:

Multiplying factor	Prefix
10^6	mega, M
10^3	kilo, k
10^{-3}	milli, m
10^{-6}	micro, μ
10^{-9}	nano, n
10^{-12}	pico, p
10^{-15}	femto, f

Historical accident prevents the wholesale use of this sub-unit system. The kilogram is the unit of mass, so that the mkg (millikilogram) is called simply the gram. Then the prefixes are used with reference to grams because it seems slightly ridiculous to put prefixes to kg. On the other hand, the kkg (kilokilogram) carries the description 'the tonne'. So far as time is concerned, the hour is 3600 seconds and 24 hours make up a day. An additional, commonly used measurement of length is the centimetre, cm, which is one hundredth of a metre ($= 0.01$ m).

Once the units of mass, length, and time have been established we can construct units for derived quantities. Sometimes these are expressed simply: often they carry a name which is a shorthand notation, sometimes regarded as part of the obscurantist jargon of technology. The quantities listed in the following table are discussed in the chapters listed in the right-hand column:

Quantity	Units	Name	Chapter
Velocity	$m\,s^{-1}$	metres per second	2
Acceleration	$m\,s^{-2}$	metres per second per second	3
Momentum	$kg\,m\,s^{-1}$	newton second, N s	5
Force	$kg\,m\,s^{-2}$	newton, N	5
Work, energy	$kg\,m^2\,s^{-2}$	newton metre, N m, or J	5
Torque	$kg\,m^2\,s^{-2}$	newton metre, N m	10
Angular momentum	$kg\,m^2\,s^{-1}$	newton metre second, N m s, or Joule second, J s	10

There are, of course, variations of these units to suit the circumstances of particular situations. The speed of a car is expressed usually in $km\,h^{-1}$, which is often written as kph (kilometres per hour), and so on.

In addition, angular measure is made in degrees (°), or radians, both of which are dimensionless: 2π radians is identical to 360°.

1

Position vectors

1.1 Position coordinates

It's not really surprising that to define a position in a one-dimensional system we need one parameter. As an example, the graduations on a metre rule are each characterized by a single number. In two dimensions we need two parameters, and three in three dimensions. If we are travelling in a railway carriage on a straight line track over a flat desert, the numbered kilometre marker posts situated at the side of the track tell us where we are in relation to the end (at 0 km) of the line and our distance from that end of the line. Let's retain the fiction that the railway track runs along the flat (the unfulfilled ambition of every railway engineer is to avoid gradients completely) but is no longer in the desert. In this engineers' paradise, however, it may not be sufficient for the railway line to be always straight. It may be cheaper and more convenient, for example, to curve the line round the edge of a lake rather than to drive it straight across on a bridge or causeway. The marker post description that we used for the straight line track is no longer adequate—the marker posts will tell us how far along the track we are from the origin, but only in the case of the straight track will this be the shortest distance of the railway carriage from the origin. Because the problem of the position of the railway carriage has become two-dimensional, we need two parameters, say x and y, each measured in km, to define the position of the carriage. For simplicity we once more choose the end of the track as origin. The line representing x, the x-axis, is then drawn from the origin so that it joins the two ends of the railway line. The line representing y, the y-axis, is then drawn perpendicular to the x-axis. It's usually convenient to have the origin of the y-axis at the same position as the origin of the x-axis, the point O. In this case, the origin is chosen to be at the position $x = 0$, $y = 0$, or more briefly, at the point $(0,0)$. If we view the railway line from a great altitude it will appear to be a single line, as shown in Fig. 1.1. The carriage will

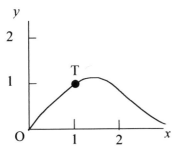

Fig. 1.1 The two-dimensional railway line with a carriage situated at the point T on the line.

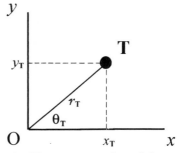

Fig. 1.2 The representation of the position of the carriage in terms of r_T.

appear as a spot on the line at the position T in Fig. 1.1, situated at the point (1, 1), i.e. at the point $x = 1$, $y = 1$.

These two parameters, which are frequently called the coordinates of T, are sufficient to define the position of T, but they need arithmetic manipulation (and we need to know the scale, what distance corresponds to an x or a y interval of 1), before we can make a statement about the distance, OT, of the carriage from O. This can no longer be described by a single parameter x, nor by the single parameter y, but it could be described by the parameter r_T, as shown in Fig. 1.2. In general, as shown in Fig. 1.2, we can say that the coordinates of the carriage's position (the point T) are x_T and y_T, so that the carriage is situated at the point (x_T, y_T). Because x_T and y_T are measured along axes which are perpendicular to each other, we may apply Pythagoras' theorem to calculate r_T in terms of x_T and y_T, as

$$r_T = \sqrt{(x_T^2 + y_T^2)}.$$

The quantity r_T alone tells us only the distance of the carriage from the origin, so that it could be situated anywhere along the circle of radius r_T passing through the (x, y) points $(0, r_T)$ and $(r_T, 0)$. Such a circle is centred at the origin which is the 'other end' of r_T. If we require a definition of the position of the carriage we need a second parameter to specify the direction in which r_T is pointing. This parameter is the angle θ_T shown in Fig. 1.2, the inclination of r_T away from the x-axis in an anticlockwise direction. From the trigonometry of a right angled triangle we know that $\tan \theta_T = (y_T/x_T)$, or that $\theta_T = \tan^{-1}(y_T/x_T)$, which is sometimes written $\theta_T = \arctan(y_T/x_T)$. Two parameters are required to define the position of the railway carriage; its distance from the origin may be deduced from, or given by, these. The polar, (r, θ), representation of the plane seems more convenient in use than the Cartesian (x, y) representation, because one of the parameters is simply the distance of the truck from the origin. While we have expressed r and θ in terms of x and y, we could equally well have given x and y in terms of r and θ, as

$$x = r \cos \theta, \qquad y = r \sin \theta.$$

If the problem is made three-dimensional, to represent a real railway line, then we may use the three parameters (x, y, z) to define the position of the carriage, or we might use spherical polar parameters (r, ϕ, θ). Conventionally, ϕ defines the orientation of the projection of r into the x-y plane and θ the orientation of r with respect to the z-axis. In this case we have

$$x = r \sin \theta \cos \phi, \qquad y = r \sin \theta \sin \phi, \qquad z = r \cos \theta$$

and the corresponding inverse relationships

$$r = \sqrt{(x^2 + y^2 + z^2)}, \qquad \theta = \tan^{-1}\left[\{\sqrt{(x^2 + y^2)}\}/z \right], \qquad \phi = \tan^{-1}(y/x).$$

Alternatively, cylindrical polar coordinates, (ρ, ϕ, z) could be used, in which case

$$x = \rho \cos \phi, \qquad y = \rho \sin \phi, \qquad z = z$$

and

$$\rho = (x^2 + y^2)^{1/2}, \qquad \phi = \tan^{-1}(y/x), \qquad z = z.$$

1.2   Position vectors

The conclusion to be drawn from this discussion is that life in one dimension is fairly easy, but that in two or more dimensions it becomes more awkward because of the proliferation of parameters. There is a symbolic method of reducing the complications introduced by the extra variables; we designate the line OT in Fig. 1.2 as a *vector* quantity which has length and direction (in the case given above the direction is defined by θ) and represent it by **r**. This is a shorthand notation which, in two dimensions, informs us that the point at issue, the one at the end of the line **r**, is distant r from the origin along the direction aligned at an angle θ to the x-axis. The length r is called the magnitude of the vector **r**, often represented symbolically by $r = |r|$. This magnitude has no direction associated with it; it is a statement simply of length, and so is not a vector quantity. To distinguish such non-vector quantities from vectors, we describe them as *scalar* quantities. If we wished to avoid defining the x- and y-directions, we might say that **r** is a vector which is formed by multiplying a vector **u**, of unit length,* in the direction of **r** by the magnitude of **r**; i.e.

$$r = ru = |r|\, u,$$

which tells us that a vector is simply the result of multiplying a unit vector by a scalar quantity. In this particular case we refer to **u** as the unit vector along the direction of **r**. A unit vector defines the unit of length along a particular direction —usually, but not always, along one of the axes in a specified set of axes.

We might equally well have used the parameters x and y to describe the position of the point at the end of the line **r**. If we move from the origin a distance x along the x-axis and then a distance y parallel to the y-axis, we find ourselves at the position defined by **r**. The vector **r** can be written as the sum of vectors x**i** in the x-direction and y**j** in the y-direction:

$$r = xi + yj$$

with **i** and **j** as unit vectors in the x- and y-directions. If we omit the unit vectors **i** and **j** in the x- and y-directions respectively, is equivalent to

$$r = x + y,$$

an equation which tells us that moving a distance x along the x-axis and then a distance y parallel to the y-axis leaves us at the position defined by **r**. In this case, x and y are called the components of **r** in the x- and y-directions. Because of the definition of the axes, **i** and **j** are at right angles to one another and so, once again using Pythagoras' theorem,

$$r = |r| = |xi + yj| = (x^2 + y^2)^{1/2}.$$

---

* This unit vector is simply the equivalent of the distance (along the direction of the line) between the markers used in the discussion of the straight railway line.

1.3   Vector addition and scalar product

To examine the process of addition of vectors, let's say that we have two vectors which we may call r_1 and r_2. What is their sum, and what might their difference be? To begin with, let's examine the case in which both r_1 and r_2 lie along the x-axis, i.e. $r_1 = x_1 \mathbf{j}$ and $r_2 = x_2 \mathbf{j}$. Summing the r's can be visualized, first of all, as the process of moving a distance x_1 along the x-axis and then moving a further distance x_2 along the x-axis. The total distance moved along the x-axis is then $(x_1 + x_2)$, so that the vector sum, R of r_1 and r_2, is given by $R = r_1 + r_2 = x_1 \mathbf{i} + x_2 \mathbf{i} = (x_1 + x_2)\mathbf{i}$. In the same way, if the two vectors to be added are both directed along the y-axis, we would obtain the result that $R = (y_1 + y_2)\mathbf{j}$. For both of these cases we have, in effect, formed the sum of two vectors along the same line by laying the vectors end to end along the appropriate axis. Then it's fairly obvious that the components of the resultant vector are $(x_1 + x_2)$ or $(y_1 + y_2)$, simply the sums of the components of the r_1 and r_2 along the appropriate axis. This observation gives us a lead into one method of adding together two vectors, for we can always say that a vector may be represented by the (vector) sum of its components along the axes. If we have two vectors r_1 and r_2, each of which has both x- and y-components so that

$$r_1 = x_1 \mathbf{i} + y_1 \mathbf{j} \quad \text{and} \quad r_2 = x_2 \mathbf{i} + y_2 \mathbf{j},$$

then we can add the components separately to give the sum, R, of the two vectors as

$$R = r_1 + r_2 = (x_1 + x_2)\mathbf{i} + (y_1 + y_2)\mathbf{j}.$$

Since we have expressed R in terms of its components, we can use Pythagoras' theorem to establish the magnitude of R as

$$|R| = |r_1 + r_2| = \left[(x_1 + x_2)^2 + (y_1 + y_2)^2\right]^{1/2}$$

and R is directed at an angle θ to the x-axis, with $\theta = \tan^{-1}[(y_1 + y_2)/(x_1 + x_2)]$. Such a vector sum is illustrated in Fig. 1.3, whence it is seen that the sum of the two vectors (their resultant, R) forms the third side of a triangle, the other sides of which are r_1 and r_2.

An alternative way of expressing the sum of two vectors can be obtained trigonometrically by using the cosine rule, which expresses the length of the third side of a triangle in terms of the lengths of the two other sides and the cosine of the angle between them. The two vectors are shown in Fig. 1.4 as r_1 and r_2, and the angle between them is given as α. Because we can write the sum of the two vectors as

$$r_1 + r_2 = R = \mathbf{i}(x_1 + x_2) + \mathbf{j}(y_1 + y_2)$$

we have once again in effect moved the 'blunt' end of the vector r_2 from the origin to the arrowed end of r_1 (or vice versa, since, e.g., $(x_1 + x_2) = (x_2 + x_1)$). When one of the vectors is moved from the origin in this way, as shown for both cases by the unarrowed full lines in Fig. 1.4, the angle between the two vectors has been altered

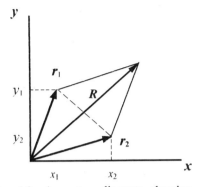
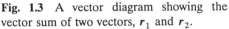

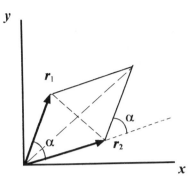

Fig. 1.3 A vector diagram showing the vector sum of two vectors, r_1 and r_2.

Fig. 1.4 The addition of two vectors to form their sum (using the cosine rule).

from α to $(180 - \alpha)$. The sum of the two vectors is thus a vector with x- and y-components $(x_1 + x_2)$ and $(y_1 + y_2)$, which forms the long diagonal of the parallelogram shown in Fig. 1.4 as a dashed line (the sum of the vectors is not necessarily the long diagonal of the parallelogram; it is the long diagonal if $\alpha < 90°$ and the short diagonal if $\alpha > 90°$). If we were dealing with the parallelogram as a geometrical object, by using the cosine rule we could write the expression for the length, R, of the long diagonal as

$$R^2 = r_1^2 + r_2^2 - 2r_1 r_2 \cos(180 - \alpha) = r_1^2 + r_2^2 + 2r_1 r_2 \cos \alpha.$$

This equation is quite straightforward geometrically, but here we are interested in the vector form and, because the expression involves products, we are led to enquire about the form of multiplication of vectors which is appropriate. We can make a beginning by examining the case of a one-dimensional vector, say, $r = x\mathbf{i}$. The magnitude of the vector is x and the question to which we need an answer is as follows: 'By what procedure can we progress from the equation defining r, $r = x\mathbf{i}$, to the equation $|r|^2 = r^2 = x^2$ or, equivalently, to $|r| = r = x$?' The answer is that we propose a form of multiplication of vectors which we symbolize in this case by a full-stop (or period) separating the two vectors. The multiplication has to satisfy the condition that $r^2 = r \cdot r = x\mathbf{i} \cdot x\mathbf{i} = x^2(\mathbf{i} \cdot \mathbf{i}) = x^2$. This requires that the vector multiplication of the unit vector $\mathbf{i}$ by itself has the value of unity: $\mathbf{i} \cdot \mathbf{i} = 1$. If the vector were directed along the y-axis a similar multiplication process would require that $\mathbf{j} \cdot \mathbf{j} = 1$. What, then, if the vector has the form $r = x\mathbf{i} + y\mathbf{j}$? We have seen that Pythagoras' theorem gives us the magnitude of the vector as $r^2 = x^2 + y^2$, so that the product $r \cdot r = r^2 = (x\mathbf{i} + y\mathbf{j}) \cdot (x\mathbf{i} + y\mathbf{j})$ must be equal to $(x^2 + y^2)$. Because $\mathbf{i} \cdot \mathbf{i} = \mathbf{j} \cdot \mathbf{j} = 1$, this equation will be satisfied only if $\mathbf{i} \cdot \mathbf{j} = \mathbf{j} \cdot \mathbf{i} = 0$. Since the angle between the two $\mathbf{i}$'s is zero and the angle between $\mathbf{i}$ and $\mathbf{j}$ is 90°, we propose that when two vectors are multiplied together in this way the result is proportional to the cosine of the angle, α, between the vectors and to the product of the magnitudes of the vectors; that is,

$$r_1 \cdot r_2 = r_1 r_2 \cos \alpha.$$

Multiplication of two vectors in this way gives a product which is a scalar quantity, and this form of multiplication is called the *scalar multiplication* of vectors.

Having defined scalar multiplication in this way we can return to our trigonometric expression for the square of the sum of two vectors and rewrite it as

$$R^2 = |R|^2 = R \cdot R = r_1^2 + r_2^2 + 2r_1 \cdot r_2$$

$$= |r_1|^2 + |r_2|^2 + 2|r_1| \, \| \, r_2| \cos \alpha$$

$$= r_1^2 + r_2^2 - 2r_1 r_2 \cos(180 - \alpha).$$

The magnitude of the difference between the two vectors, the length, R', of the diagonal joining the arrowed ends of the vectors r_1 and r_2 in Fig. 1.4 (the short diagonal in Fig. 1.4), is given by

$$(R') = |R'|^2 = |r_1 - r_2|^2 = |r_1|^2 + |r_2|^2 - 2r_1 \cdot r_2$$

$$= r_1^2 + r_2^2 - 2r_1 r_2 \cos \alpha.$$

Once the value of R (or R') has been established, we can use the sine rule to find the other angles, say β and γ, in the triangles; for example,

$$\frac{R}{\sin \alpha} = \frac{r_1}{\sin \beta} = \frac{r_2}{\sin \gamma},$$

so that the triangle (or parallelogram) formed by the addition of the two vectors is completely specified.

The addition or subtraction of two vectors may be performed arithmetically by using the technique of adding together components or by using trigonometric techniques, as outlined above. Such a calculation could also be carried out by a graphical method. The vectors are drawn to scale and the other sides of the parallelogram are drawn in. Then the lengths and directions of the diagonals define the vectors which are the sum and difference of the two vectors. Measurement of the diagonal lengths and of the angles in the parallelogram specifies the sum and/or difference of the two vectors. As an alternative, two different vector triangles, those which are formed from a diagonal and two sides of the parallelogram, could be used to represent the sum or difference of the two vectors.

Example 1. A fishing boat leaves its home port and travels 24 km in a northeasterly direction before it drops anchor and begins fishing. A second vessel, the master of which is more experienced in the ways of the fish in the local fishing ground, sets sail one hour later in a direction south by south-east. This second boat drops anchor to begin fishing after it has travelled 36 km. How far apart are the two boats when their crews are fishing?

Even though the sea often gives the impression that it is anything but flat, this problem is two-dimensional. Local variations of elevation do not affect, for example, the distance between two different ports. So far as we are concerned, any

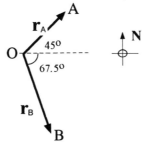

Fig. 1.5 The positions and tracks of the two fishing boats, A and B.

motion of the boats in the third dimension has no effect on the distance they have travelled—we may deal with this example by adopting the convenience that the sea is always flat and calm. This leads us to represent the points on an *xy*-plane, and their tracks as vectors joining those points to the origin, the port, shown as O in Fig. 1.5. The *x*-axis is chosen conventionally as the E–W line and the *y*-axis as the N–S line. The direction NE in terms of the angle used in the (r, θ) representation of a plane is 45° and of SSE is $-67.5°$ (or 292.5°), and the *r*'s have values of 24 km and 36 km respectively.

The most obvious way of finding the separation of the two craft is to make a scale drawing, as in Fig. 1.5, draw the line AB, and measure its length, a standard navigational practice. Here we approach the problem trigonometrically. From the cosine rule for a triangle with two sides of length 24 and 36 which include an angle of 112.5° between them, we may write

$$AB^2 = 24^2 + 36^2 - (2 \times 24 \times 36)\cos 112.5°$$
$$= 24^2 + 36^2 + (2 \times 24 \times 36)\cos 67.5°$$

which gives the result

$$AB = 50.3 \text{ km}.$$

If we needed to know the heading along which the vessel at A should travel to benefit from the more profitable fishing waters at the position of B, we would need to find out the angle A in the triangle OAB. From the sine rule for triangles,

$$\frac{OB}{\sin A} = \frac{AB}{\sin 112.5°},$$

giving

$$\sin A = \frac{36 \sin 112.5°}{50.3}$$

or $A = 41.4°$. The angle at which the line AB crosses the *x*-axis is therefore $[180 - (45 + 41.4)]° = 93.6°$. The course to be steered from A to B by the helmsman is 3.6° west of south. It's unlikely that an unsophisticated craft such as a fishing boat could be steered with that degree of precision. Most likely it would follow a bearing due south until vessel B hove into sight and then its course would be adjusted suitably.

Example 2. An aircraft flies from an airfield at town C in a north-westerly direction until it passes over town D, which is 150 km from C. It then changes course to head towards its destination at town F, which is 100 km due west of D. How much further has the aircraft travelled than it would have done by flying directly from C to F? What is the bearing of F from C?

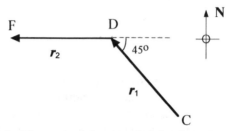

Fig. 1.6   The track of the aircraft flying from C to F via D.

The diagram for this problem is given in Fig. 1.6. The distance travelled by the aircraft in flying along the path CDF is 250 km. The vector CF is the vector sum of CD and DF; in other words,

$$r = r_1 + r_2.$$

The angle included between the two vectors r_1 and r_2 is 135°, so that

$$r^2 = 100^2 + 150^2 + (2 \times 100 \times 150)\cos 45°,$$

which gives the magnitude of CF as $r = 231.76$ km. The difference in distance between the direct path and the course followed is only 18.2 km, a mere 7%. The advantage of 'cutting the corner' is not as large as might be expected.

The bearing of F from C is obtained by finding the angle DCF from the sine rule:

$$\frac{100}{\sin \text{DCF}} = \frac{231.8}{\sin 135°}, \qquad \text{DCF} = 17.8°.$$

The bearing of F and C is therefore 62.8° W of N, or 27.2° N of W.

Example 3. Captain Ahab has been informed by the crew of a passing whaler that Moby Dick has been sighted 300 km away in a south-westerly direction. The *Pequod*'s whale boats were all splintered to matchwood during the last encounter with a school of whales and the ship must call in at Valparaiso, some 100 km SSE of its present position, to pick up some planks to make more boats. In what direction should Ahab set course from Valparaiso to reach the point at which the white whale was sighted? How far will the *Pequod* sail after leaving the port before it arrives at the sighting place?

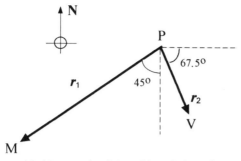

Fig. 1.7 An extract from Ahab's pursuit of the white whale. P is the position of the *Pequod*, M the position at which Moby Dick was sighted, and V the port of Valparaiso.

In this problem the track of the ship takes it to the port, V in Fig. 1.7, and then to the location of the sighting, M. The sum of these two tracks must be equivalent to the direct course from the ship's present position, P, to M. If the unknown course VM is designated as r, then $r_2 + r = r_1$, so that

$$r = r_1 - r_2.$$

The angle between r_1 and r_2 is 67.5°, so that

$$r^2 = 300^2 + 100^2 - (2 \times 100 \times 300)\cos 67.5°,$$

giving $r = 277.6$ km.

The course to be steered is found by calculating the angle PVM,

$$\frac{300}{\sin \text{PVM}} = \frac{277.6}{\sin 67.5°}, \qquad \text{PVM} = 86.9° \text{ or } 93.1°,$$

an ambiguity which comes about because $\sin \theta = \sin(180 - \theta)$. Very often, the choice between these two possibilities is obvious, but in this case it is not. In order to select the correct value of PVM it is necessary to use the sine rule to calculate the angle PMV, from

$$\text{PMV} = \sin^{-1}[(100 \sin 67.5°)/277.6] = 19.48°.$$

Since the sum of the angles in the triangle PMV must be 180°, the angle PVM must be 93.1°. The course to be sailed is 25.6° S of W.

There is no reason why we could not have solved these problems by writing down the position vectors in terms of their x- and y-components and the unit vectors along the E–W and N–S directions. Then the process of addition or subtraction of components in the different directions would have led to the components of the resultant vector, which would have been used to calculate the magnitude and direction of the resultant vector. Ahab himself would have used a graphical method. Such methods are just as valid as the trigonometric method used in the preceding examples.

Exercises

1.1. Taking the N–S and W–E lines as the y- and x-axes respectively, solve the three examples (1, 2, and 3) given above by adding their x- and y-coordinates to find the resultant vector and its direction.

1.2. Taking one of the vectors, say r_1, as the x-axis, solve the three problems given above in terms of the components of the vectors along this new x-axis and the y-axis perpendicular to it.

Solutions

1.1. *Example 1:* The vector joining A and B, r, is obtained from $r_A = r_B + r$, i.e.

$$r = r_A - r_B$$

with

$$r_A = i(24\cos 45°) + j(24\sin 45°) = 16.97i + 16.97j$$

and

$$r_B = i(36\cos 67.5°) - j(36\sin 67.5°) = 13.78i - 33.26j$$

so that

$$r = 3.19i + 50.23j.$$

Then

$$r = |r| = 50.3, \qquad \theta = \tan^{-1}(50.23/3.19) = 86.4° \text{ or } 266.6°.$$

The ambiguity in θ arising because $\tan\theta = \tan(180 + \theta)$ gives the course to be steered from B to A or from A to B. These two courses are N3.6°E or S3.6°W.

Example 2:

$$r = ir_x + jr_y$$

with

$$r_x = r_x(CD) + r_x(DF) \qquad \text{and} \qquad r_y = r_y(CD),$$

where $r_x(CD)$ is the x-component of CD, etc.

$$r = i206.6 + j106.6, \qquad |r| = 231.8\,\text{km}, \qquad \theta = \text{W27.2°N}.$$

Example 3:

$$r_1 = 38.3i - 92.4j, \qquad r_2 = -212.1i - 212.1j,$$

$$r_1 - r_2 = 250.4i + 119.7j,$$

$$|r| = 277.6\,\text{km}, \qquad \theta = \text{W25.6°S}.$$

1.2. *Example 1:* Taking NW as the *x*-axis,

$$r_1 = i24, \qquad r_2 = -i13.78 - j33.26,$$
$$r = 37.78i + 33.26j, \qquad |r| = 50.3,$$
$$\theta = 41.36°, \qquad \text{giving a course S3.6°W.}$$

Example 2: Taking CD as the *x*-axis and C as origin,

$$r_1 = 150i, \qquad r_2 = -70.71i - 70.71j,$$
$$r = r_1 - r_2 = 220.7i - 70.71j,$$
$$|r| = 231.76, \qquad \theta = 17.76°, \quad \text{i.e. N62.8°W.}$$

Example 3: Taking PM as the *x*-axis,

$$r_1 = 38.27i + 92.35j, \qquad r_2 = 300i,$$
$$r_2 - r_1 = 261.73i - 92.38j,$$
$$|r| = 277.6 \, \text{km}, \qquad \theta = 19.44°, \qquad \text{i.e., W25.6° S.}$$

2
Velocity vectors

2.1 Velocity

The concept of position vector that we have developed so far seems obvious. Only when we introduce the ideas of axes and components do we move away from the simplest commonsense ideas. But common sense tells us that in many cases bodies move about—their position changes with time—so that we have the task of inventing some form of time dependent position vector. Otherwise, the representation of position by a vector will be of quite limited use.

In one dimension, if our railway carriage (in Chapter 1) increases its displacement from the origin at O by $(x_2 - x_1)$ m in the time interval separating the times t_1 and t_2 (measured in seconds), we can say that the average velocity V of the carriage in that time interval is

$$V = \frac{(x_2 - x_1)}{(t_2 - t_1)} \ \mathrm{m\,s^{-1}}.$$

We may illustrate this relation by setting up a coordinate system with x, position, as vertical axis and t, time, as horizontal axis, as shown in Fig. 2.1. We draw a line BC joining the points the coordinates of which are (t_1, x_1) and (t_2, x_2) respectively and complete a triangle BCD by drawing the lines BD and CD parallel to the axes. Because this triangle is right angled, the ratio $(\mathrm{CD}/\mathrm{BD}) = (\tan \theta)$ is called the gradient of the line BC. The lines CD and BD on the graph represent the numerator and denominator in the expression for the average velocity, V, so that the gradient of the straight line BC, the tangent of the angle θ in Fig. 2.1, is the

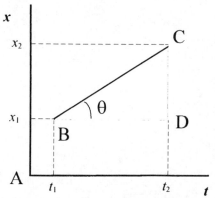

Fig. 2.1 A graph of displacement, x, against time, t, showing how the average velocity, V, is constructed.

average velocity. In reality, the position of the carriage is unlikely to vary uniformly with time—signals, intervening stations, gradients in the line, and curves on the track are all likely to make the displacement–time graph vary markedly from the straight line joining B and C in Fig. 2.1. In this real case the line on the graph representing position as a function of time begins at B and still ends at C, but the line joining them is no longer straight. The average velocity remains the same, but that does not mean that the velocity at all times, or even almost at any time, between t_1 and t_2 is the same as the average velocity. This observation makes us enquire as to how we might define the velocity of the carriage at every instant of time during its journey from B to C. The answer is by differentiation; we draw a triangle in Fig. 2.1 which is a microscopic analogue of the triangle BCD. In this new triangle the time interval, from t to $t + \delta t$ and the change in position from x to $x + \delta x$ are both very small.* The hypotenuse of this triangle joins together two points on the position–time curve the separation of which is so small that between them the curve is represented effectively by a straight line. The velocity during the time interval δt is given by $V = \delta x / \delta t$, an average velocity because the curve and hypotenuse of the triangle are not an exact match. In the limit as $\delta t \to 0$ the quantity

$$\underset{\delta t \to 0}{\text{limit}} \left(\frac{\delta x}{\delta t} \right)$$

is the gradient of the curve at the point on the curve under consideration. While $\delta x / \delta t$ was an average velocity its limiting value is an instantaneous velocity, corresponding to the gradient of the displacement-time curve at a particular value of t (or of x). We now call the instantaneous velocity (the actual velocity for that value of t) simply the velocity v so that we can write

$$v = \underset{\delta t \to 0}{\text{limit}} \left(\frac{\delta x}{\delta t} \right) = \frac{\mathrm{d}x}{\mathrm{d}t},$$

an expression in which $\mathrm{d}x/\mathrm{d}t$ represents the limiting value of $(\delta x / \delta t)$ as $\delta t \to 0$. $\mathrm{d}x/\mathrm{d}t$ is called the first derivative of the displacement, x, with respect to time, t.

2.2   Velocity as a vector

This definition of velocity seems quite satisfactory until we examine it more closely. What we have omitted from our discussion so far is that the direction of travel is included implicitly in the relation involving v (or V) and x. This comes about because x may increase or decrease during the journey, depending on whether the carriage travels from B to C or from C to B. Whichever of these directions is taken, time is always increasing. It is quite legitimate to consider a negative interval of time, so that events of the past may be located in time. If this were not allowed, it would be impossible to give precise descriptions of yesterday's happenings, let alone of occurrences from a long time ago. The parameter time, on the other hand,

---

* The convention is that these very small changes, in time and in position, are represented by δt and δx respectively.

always increases. 'Time's winged chariot hurrying near' goes along a one-way street, in the direction of increasing time. The old saying 'You can't turn the clock back' is a verbal statement that time changes unidirectionally. Time travel can happen only in the mind. This means that if a return journey is made by the carriage the trip from C to B must follow the one from B to C. The line on the graph describing the return leg of the journey will have an average slope which is of opposite sign from that for the outward leg of the journey (but not necessarily of the same magnitude). This happens because in the one case x increases as t increases, while in the other x decreases as t increases further. The directionality of x, implicit in the definition of an origin for the coordinate system, is reflected by a directionality of the quantity v, the velocity. So far we have worked in one dimension and have neglected to use the x unit vector to define the direction of x, because the unit vector seems redundant in one dimension. Strictly, to remain consistent with our vector ideas, we should write $x = x\mathbf{i}$, with $\mathbf{i}$ the unit vector in the x-direction, so that the expression for the velocity should be written as $d(x\mathbf{i})/dt$. Since the unit vector remains unchanged with time—that is, we have chosen a fixed direction for our x-axis and the magnitude of $\mathbf{i}$ is fixed—we may write the velocity in vector form (for this one-dimensional case) as

$$v = \frac{d(x\mathbf{i})}{dt} = \mathbf{i}\frac{dx}{dt} = \mathbf{i}v_x.$$

Velocity is a vector quantity, the components of which in, for example, two dimensions, v_x and v_y, are defined as the rates of change of position in the x- and y-directions, so that

$$v = \mathbf{i}v_x + \mathbf{j}v_y = \mathbf{i}\frac{dx}{dt} + \mathbf{j}\frac{dy}{dt}.$$

The form of this relationship encourages us to use v_x and v_y as the coordinates of a two-dimensional coordinate system, as shown in Fig. 2.2. The point P, (v_x, v_y), in the coordinate system is at distance

$$v = \left(v_x^2 + v_y^2\right)^{1/2}$$

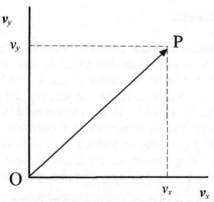

Fig. 2.2 The velocity vector, v, drawn in the v_x–v_y velocity plane.

from the origin, O, and the line OP has a gradient $\tan \theta = v_y/v_x$. The line OP is the vector $\boldsymbol{v}$, the velocity vector, the magnitude of which is v. Often, v is called the speed associated with the velocity $\boldsymbol{v}$. v is a scalar quantity, having no direction associated with it. Thus we might be able to say, for the railway journey from B to C and return, that the speeds in either direction are the same—in such a case the velocities on the two legs of the journey would be equal in magnitude but opposite in sign.

One point to note is that the velocity vector is drawn on a graph, in this case in two dimensions, which has axes described by the velocity components, v_x and v_y, in the x- and y-directions. The analogous graph in three dimensions, v_x, v_y, and v_z, describes a 'space' in which 'lengths' are measured in terms of velocity. Such a space is described as a velocity space.

2.3 Distance–time relations

From the definition of the average velocity it is obvious that we could write

$$x_2 - x_1 = V(t_2 - t_1)$$

or

$$x_2 = V(t_2 - t_1) + x_1$$

which, if we set $t_1 = 0$ and write $t_2 = t$, becomes

$$x_2 = Vt + x_1,$$

where V $(= \tan \theta)$ is the gradient of the straight line BC drawn on the displacement–time graph (Fig. 2.1). When we look at the process of expressing x in terms of t in the case in which the velocity is the instantaneous velocity, dx/dt, it's no longer sufficient to multiply one side of the equation by the denominator of the other. The quantity to be multiplied, v, is not necessarily constant, and in practice is most unlikely to be always constant in the time interval from t_1 to t_2. We take v expressed in derivative form and subject it to a procedure which is the equivalent of that just used to find x_2. In the one-dimensional case (which allows us to drop the vector notation) this produces the relation between a small change in position, δx, and a small increase in time, δt,

$$v \, \delta t = \delta x.$$

If the time interval in which we are interested is finite, say $t_2 - t_1$, we have to add up every $v \, \delta t$ contribution, one from each of the δt's which make up $t_2 - t_1$. In the limit as $\delta t \to 0$ the total obtained is the equivalent of adding up all the infinitesimal contributions dx to the displacement which takes place in the interval $t_2 - t_1$. This process of addition cannot be written as a simple sum of discrete terms, because dx and dt are both infinitesimally small, and the sum is of infinitesimal quantities which vary continuously. To deal with this we use the procedure of integration. We write this integration (using $\int$ as the integration sign) between the limits t_1 and t_2 as

$$\int_{t_1}^{t_2} v \, dt = \int_{t_1}^{t_2} \frac{dx}{dt} \, dt = \int_{x_1}^{x_2} dx.$$

The integration of dx is between the limits x_1 and x_2 which correspond with the t limits of the time dependent integral. If v is independent of t, that is, it is a constant velocity or an average velocity, V, then since

$$\int_{x_1}^{x_2} dx = x_2 - x_1 \qquad \text{and} \qquad \int_{t_1}^{t_2} dt = t_2 - t_1$$

we have $V = (x_2 - x_1)/(t_2 - t_1)$, which was our basic definition of average velocity.

The proper completion of the integration process requires that three of the limits t_1, t_2, x_1, and x_2 are defined for the integrals. When this condition is satisfied we can say we know the boundary conditions that define the problem.

2.3.1 Boundary conditions

Example 4. A hiker walks with a constant velocity $v = 3\,\mathrm{m\,s^{-1}}$. If his displacement at a time chosen as $t = 0$ is $1\,\mathrm{m}$, what is his position at the subsequent time $t = 3\,\mathrm{s}$?

The boundary condition here is that at $t = 0$, $x = 1$, and the third limit, $t = 3$, is given. Using the method in the previous paragraph,

$$\int v\,dt = \int dx$$

or

$$\int_0^3 3\,dt = 3 \int_0^3 dt = \int_1^x dx,$$

which is evaluated as

$$3|t|_0^3 = 3(3 - 0) = 9 = x - 1,$$

giving

$$x = 10\,\mathrm{m}.$$

Equally, we could have performed the integration without limits by saying, in effect, that x represents the displacement after $t = 0$. In this case the equation of integration would have been written as

$$\int v\,dt = \int dx.$$

Because v is constant, independent of time, this equation is solved by

$$vt + C = x,$$

where C is the constant of integration. The constant is evaluated by substituting into the solution the boundary condition $x = 1$ for $t = 0$, giving

$$C = 1$$

so that

$$x = vt + 1$$

which at $t = 3$ has the value $x = 10\,\mathrm{m}$.

Boundary conditions, which may be expressed as limits of integration, are required to make it possible to complete the integration properly. Had we mistakenly omitted the constant of integration, C, from the second solution of this problem, we would have obtained the answer $x = 9\,\text{m}$, which is the displacement that has taken place in the three seconds but is not the position (the complete displacement).

In this example, the velocity was constant, so that the velocity and the average velocity were the same. We might as well have substituted the numbers in our definition of the average velocity to find the position at $t = 3$. Obviously, that would have been exactly the same as the integration between limits given in the solution of the problem.

2.3.2 Relative position and relative velocity

Example 5. In a stock car race along a straight course, car number 1 has its velocity described by a vector v, while car number 10 has its velocity described by a vector $-v$, equal in magnitude but opposite in direction. Will the two cars collide?

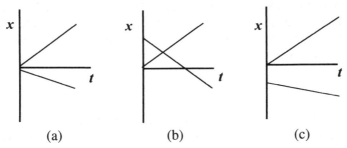

       (a)   (b)   (c)

Fig. 2.3 Different examples of displacement–time graphs for two stock-cars with equal and opposite velocities.

For convenience, choose the axis v_x so that v coincides with it. This means that we may treat the motion as one-dimensional and investigate it by means of an x–t graph. The displacements of the two cars can then be represented by two straight lines of equal and opposite slopes, giving rise to the three possible solutions, shown in Fig. 2.3. In Fig. 2.3(a) the cars coincide at the origin, at which $x = 0$ and $t = 0$, and then set off in opposite directions so that they can never collide. In Fig. 2.3(b) they have different displacements at $t = 0$, the lines cross, and their positions will coincide at the time and position defined by the crossing point. In Fig. 2.3(c) the cars again have different displacements when $t = 0$, but the lines never cross so that the positions of the cars can never coincide. Even when the two cars travel along the same straight line (their motion is collinear) they will not necessarily collide.

By drawing the x–t graphs we have, in effect, integrated the equation

$$\pm|v| = \pm v_x = \frac{\mathrm{d}x}{\mathrm{d}t}$$

so that

$$\pm \int v\,\mathrm{d}t = \pm v \int \mathrm{d}t = \int \mathrm{d}x,$$

giving

$$\pm vt + C_x = x,$$

where C_x, the constant of integration, has to be determined from the boundary conditions; for example, if $x = 0$ when $t = 0$, then $C_x = 0$. If that condition is appropriate for both cars (as illustrated in Fig. 2.3(a)), then

$$x_1 = vt \quad \text{and} \quad x_2 = -vt,$$

so that x_1 and x_2 can be equal only when $t = 0$. On the other hand, if $x_1 = 0$ and $x_2 = x_2(0)$ when $t = 0$, we have

$$x_1 = vt \quad \text{and} \quad x_2 = -vt + x_2(0),$$

as illustrated in Figs. 2.3(b) and (c). In this case $x_1 = x_2$ when $t = x_2(0)/2v$, and the two cars collide if $x_2(0)$ is positive or do not collide if $x_2(0)$ is negative.

The answer to the question posed in this example thus appears to be dependent on the boundary conditions. Here, so far, for clarity, we have considered only the velocity component v_x and have ignored the components v_y and v_z simply because they are zero in the velocity space we have defined. Take the case of v_y as an example—when this is zero the value of y associated with the body is unchanged in time so that $y(t) = y(0)$. Looked at from the point of view of the integration process, this means that

$$y = \int v_y \, dt + C_y = C_y$$

with $C_y = y(0)$. The velocities of the two cars in the y-direction are identical—both are zero—but until the boundary conditions are known we are unable to say whether or not they are travelling in the same xy-plane. Introducing the component $v_z = 0$ leads to the conclusion that without the boundary conditions we do not know whether the cars are travelling in the same xz-plane, but we do know that each travels in its own particular yz-plane. If the boundary conditions were such that these two yz-planes were the same, the vectors would be coplanar.

The conclusion to be reached from this is that the cars might collide if their velocity vectors were collinear, but even when they are collinear the cars could be moving away from, rather than towards, each other.

When we discussed position vectors, we made the observation that the origin was chosen at a convenient location, that there is no absolute position which defines the origin of a coordinate system. When, in Example 1, we calculated the positions of the two ships with respect to each other, rather than relative to their home port, we moved the origin, in effect, from the harbour to one or other of the ships. The separation and bearing of one of the boats from the other is measured from the point at which the other is situated. This separation and bearing of one boat measured relative to the other is called its position relative to that boat, or its relative position. Obviously, relative position is of primary importance to the seafarer, but only of secondary importance to the landlubber on the quayside.

The utility and convenience in using a relative position vector in this way to define position suggests that when we discuss velocity as a vector there might be profit in using the concept of relative velocity. As an illustration of this idea, recollect the very simple case of the one-dimensional railway on which a carriage is travelling with a velocity v. That is the velocity which we observe when we stand at the side of the track and see it pass us by going in a direction, say, due north. If instead of standing on the trackside we were riding in the carriage, what would our observation be? We would see, looking forwards along the track, that our destination appeared to approach us with a speed v; that is, in the direction due south. Our velocity with respect to the track is v, but the velocity of our destination (and of our departure point) relative to us is $-v$. In one case the origin of our coordinate system is fixed somewhere along the line of the track; in the other it is fixed at some position in the carriage. These two different coordinate systems are often described as two different frames of reference, and the results of our observations depend on the frame of reference from which the observation is made. Just as there is no absolutely defined position for the origin of spatial coordinates, so there is no absolutely defined origin in velocity space from which velocity is measured. In the simplest case, that of the one-dimensional railway, the consequence of changing the frame of reference from the track to the carriage was a reversal of the sign of the velocity, v, but if there were a second railway carriage travelling with a velocity v' (oppositely directed to v) relative to the track we would see it from the first carriage as travelling with velocity $-(v' + v) = -(v + v')$. Its velocity relative to the first carriage is the negative sum of the two velocities relative to the track. In the same way, if the two carriages were travelling in the same direction, the relative velocity of the second carriage observed from the first would be $v' - v$, the difference of the two velocities. The velocity of the first carriage relative to the second would be of the same magnitude but opposite in sign; that is, $-(v' - v) = (v - v')$.

In two and three dimensions the situation is rather more complicated because, in general, the sums and differences of the vectors are not collinear with the vectors.

Example 6. Bonnie and Clyde hold up a gas station $10\,\text{km}$ north of the sheriff's office in the town of Scrapeep. They make their getaway at $120\,\text{km}\,\text{h}^{-1}$ along the straight desert road, due north, heading towards the state boundary, which is $140\,\text{km}$ north of the gas station. If the sheriff in his office hears of the robbery 15 minutes after it has taken place, and immediately drives in pursuit at $160\,\text{km}\,\text{h}^{-1}$ along the same road, will he catch the criminals before they escape across the state line?

The velocities of the two vehicles are 120 and $160\,\text{km}\,\text{h}^{-1}$ due north. Their relative speed is $40\,\text{km}\,\text{h}^{-1}$. The velocity of the sheriff's vehicle relative to the getaway car is $40\,\text{km}\,\text{h}^{-1}$ (the velocity of the getaway car relative to the sheriff's is $-40\,\text{km}\,\text{h}^{-1}$). By the time the sheriff sets off in pursuit, the getaway car is $40\,\text{km}$ north of his office and he would have to travel for $1\,\text{h}$—that is, a distance of $160\,\text{km}$—before catching it up. In that time the thieves would have travelled $120\,\text{km}$ and would be $10\,\text{km}$ on the safe side of the state line (unless the highway patrol picked them up for speeding!).

More formally, we could have written

$$v_1 = \mathrm{d}x_1/\mathrm{d}t = 120\,\text{km}\,\text{h}^{-1}, \qquad v_2 = \mathrm{d}x_2/\mathrm{d}t = 160\,\text{km}\,\text{h}^{-1},$$

where x_1 and x_2 are the positions of the vehicles, and we have dropped the vector notation because the problem is one-dimensional. These equations may be integrated to give

$$x_1 = v_1 t + C_1 \quad \text{and} \quad x_2 = v_2 t + C_2,$$

where the C's are integration constants, to be determined by the boundary conditions. We may choose the zero of time as the time at which the criminals left the scene of their crime and the zero of distance as the location of the sheriff's office. Then for the first equation we have as the boundary condition $x_1 = 10$ for $t = 0$, so that

$$x_1 = 120t + 10.$$

Until $t = \frac{1}{4}$ h (the unit of time here is one hour), $x_2 = 0$, so that the boundary condition for the patrol car is $x_2 = 0$ for $t = \frac{1}{4}$, giving

$$x_2 = 160(t - \tfrac{1}{4}) = 160t - 40.$$

The positions of the vehicles will coincide when $x_1 = x_2$, or

$$120t + 10 = 160t - 40$$

or

$$t = \tfrac{5}{4} \text{h},$$

at which time

$$x_1 = 120(\tfrac{5}{4}) + 10 = 160 \,\text{km},$$

10 km beyond the state line. In this instance, crime seems to have paid.

The complication produced by more realistic examples may be illustrated by moving to a two-dimensional case in which we cannot substitute scalar quantities for the vectors.

2.4   Position coincidence and closest approach

Example 7.   A fishery protection vessel steaming N30°W at $48\,\text{km}\,\text{h}^{-1}$ observes at noon a trawler steaming due N at $20\,\text{km}\,\text{h}^{-1}$. If the two boats meet at 12:45 hrs, what were the distance and the bearing of the trawler from the position vessel at noon?

The first step here is to establish the magnitude and direction of the relative velocity. The two velocities are drawn in Fig. 2.4(a) and their representation in velocity space is shown in Fig. 2.4(b), whence, by applying the cosine rule, we find that

$$|v_r|^2 = 48^2 + 20^2 - 2 \times 20 \times 48 \cos 30°,$$

giving

$$v_r = 32.27 \,\text{km}\,\text{h}^{-1}.$$

The direction of the relative velocity is given by

$$\frac{32.27}{\sin 30°} = \frac{48}{\sin \alpha}, \quad \text{i.e. } \alpha = 131.9°,$$

so that the relative velocity is directed (nearly) N48°W or W42°N. (Without the

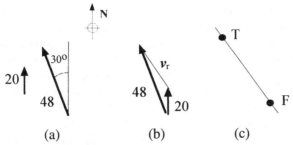

(a)   (b)   (c)

Fig. 2.4 (a) Vectors representing the velocities of the two ships. (b) The vector diagram drawn in velocity space to find the magnitude and direction of the relative velocity. (c) A representation of the track mapped out by the relative velocity, in which F and T represent the positions of the fishery protection vessel and the trawler at noon.

vector diagram of Fig. 2.4(b), it is possible that we might have found $\alpha = 48.1°$, since $\sin \alpha = \sin(180 - \alpha)$.

Once we know the direction and magnitude of the relative velocity we may draw the track of the fishery protection vessel observed from the trawler, as in Fig. 2.4(c). The two boats coincide in position when the time lapse from noon, $t = 0.75$ h, so that their separation at noon was $r = (0.75 \times 32.27) = 24.2$ km. The length of the track line after noon is $(24.2 - v_r t)$, taking TF as the positive direction and T as the origin.

It may be that the captain of the fishery protection vessel wanted to inspect the trawler's nets and so arranged the coincidence of position. For the most part, ship's masters try to avoid collisions, and so their objective is to keep their distances of closest approach to other vessels well outside a margin of safety. The distance of closest approach is visualized easily in terms of a diagram such as Fig. 2.4(c). Say that the trawler is so positioned at $t = 0$ that the track of the protection vessel takes it some distance away from the trawler. For argument's sake, let the initial bearing (at noon) of the trawler from the protection vessel be W30°N and let their initial separation be 20 km. This difference in initial positions of the two ships has no relevance to their relative velocity, the magnitude and direction of which will be exactly the same as that shown in Fig. 2.4(b). The difference comes about when we draw the relative course diagram. This time the trawler's initial position does not lie on the track traced by the relative velocity, as shown in Fig. 2.5. The line TF

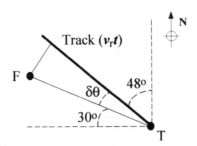

Fig. 2.5 The initial positions of the two vessels, the track of F relative to T, and the distances separating F and T at various times.

makes an angle $\delta\theta = 12°$ with the track line, and the distance between the two ships is the length of the line joining T and the point distant $v_r t$ from F along the direction of the relative velocity. Three of these lines are shown dashed in Fig. 2.5, and it's obvious that the closest approach of F to T occurs when the line of separation is a normal to the track; that is, the triangle formed in Fig. 2.5 is a right angled one. The distance of closest approach is then given simply by $d = \text{TF} \sin \delta\theta$, so that using the numbers chosen above, $d = 20 \sin 12° = 4.16 \, \text{km}$. This closest approach will occur at a time t deduced from $v_r t = \text{TF} \cos \delta\theta$, giving here $t = (20 \cos 12°)/32.26 = 0.61 \, \text{h}$. The two vessels will pass within 4.16 km of each other at 12:36 hrs.

Example 8.   A submerged submarine is travelling at $45 \, \text{km h}^{-1}$ on a course N60°E. At 08:00 hrs it is detected at a point B by a helicopter which reports to a destroyer some 30 km east of B. Assuming that the submarine captain sees no reason to change course or speed, what is the minimum speed at which the destroyer must pursue the submarine if it is to make an interception?

If the destroyer is to intercept the submarine, the relative velocity of the two ships must be due E or due W, depending on whether it is measured from the submarine's or the destroyer's frame of reference. In velocity space the velocity vectors representing the submarine's velocity and the relative velocity are arranged as shown in Fig. 2.6, and the triangle of velocities is completed by the vector

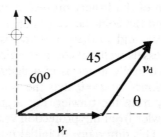

Fig. 2.6 The velocity space diagram representing the pursuit by the destroyer of the submarine.

velocity of the destroyer, v_d. Because the magnitude of the relative velocity is not specified, the course of the destroyer, Eθ°N, can take any appropriate value. The value of v_d is given by the sine rule:

$$\frac{v_d}{\sin 30°} = \frac{45}{\sin(180 - \theta)} = \frac{45}{\sin \theta}$$

or

$$v_d = \frac{45 \sin 30°}{\sin \theta} = \frac{22.5}{\sin \theta},$$

which has a minimum value for the maximum possible value of $\sin \theta$, equal to 1,

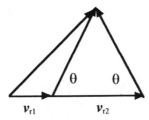

Fig. 2.7 The velocity vector diagram for interception problem when the pursuer's velocity is more than the minimum.

which occurs for $\theta = 90°$. The minimum value of v_d which will complete the triangle is therefore

$$v_d = 22.5 \, \text{km h}^{-1}$$

on a course which is due north.

This result shows that the minimum velocity for interception is in some ways analogous to the distance of closest approach. The closest approach occurs when the line of separation is normal to the track; the minimum velocity occurs when the resultant velocity is directed perpendicular to one of the velocities. This point is emphasized by the way in which the sine rule was written in the solution. Because $\sin(180 - \theta) = \sin \theta$, there is always an ambiguity in the interception problem, except when the velocity is a minimum, in which case $(180 - \theta) = \theta$ because $\theta = 90°$. We can illustrate this by examining the situation which arises when the destroyer steams at $36 \, \text{km h}^{-1}$ (well above the minimum value), giving rise to two possibilities for its course, as illustrated in Fig. 2.7, which shows there are two values of v_r which will lead to interception taking place. The values of these may be obtained from the cosine rule:

$$v_r^2 = 36^2 + 45^2 - [90 \times 36 \cos(\theta - 30)]$$

or

$$v_r^2 = 36^2 + 45^2 + [90 \times 36 \cos(180° - 30° - \theta)]$$
$$= 36^2 + 45^2 + [90 \times 36 \cos(30° + \theta)].$$

Because we know the value of v_d, the value of θ is obtained from the sine rule as

$$\theta = \sin^{-1}[(45 \times \sin 30°)/36] = 38.7°,$$

so that

$$v_r = 10.9 \, \text{km h}^{-1} \text{ or } 67.1 \, \text{km h}^{-1}$$

and the interception will take place at 10:45 hrs or at 08:27 hrs respectively. These courses could be described as a shadowing course, closing to follow the submarine directly, or an attack course, closing as quickly as possible. The shadowing course would not be particularly useful for carrying out an attack because only a submarine captain with a well developed death wish would continue on the same course for more than two hours. In reality, it's unlikely that he would maintain his course even for the 27 minutes it would take for the destroyer on an attack course to intercept him.

Exercises

2.1. A man standing conversing with a friend during a downpour of rain finds that he obtains maximum protection from the rain when he holds his umbrella vertically upwards.* When he leaves his friend and sets off home at $5\,\mathrm{km\,h^{-1}}$ he holds his umbrella at an angle of $10°$ to the vertical for best protection. What is the velocity of the rainfall?

2.2. A frigate sails at 10 knots ($1\,\mathrm{knot} = 0.57\,\mathrm{m\,s^{-1}}$) close to a hostile coast and will pass within range of a shore-based battery of cannon. If the cannonballs fired from the battery have a velocity of $280\,\mathrm{m\,s^{-1}}$, what angular allowance must the gunners make to their gunsights if they wish to hit the frigate when it is nearest to the battery?

2.3. A boy who can swim with a velocity u in still water goes bathing in a river which flows with a velocity v. He swims 50 m in a direction perpendicular to the river bank to a platform moored in the river and, after sunbathing on the platform for a while, decides to cool himself down by swimming 50 m upstream and back to the platform. At lunchtime he returns to the bank at the point at which he entered the water, taking the direct route again. What is the ratio of the time that he has spent in the water going from the bank to the platform and back and the time that he spent in his refreshing swim along the river? What would happen if $v > u$?

2.4. An escaped convict needs to cross a stream of width a from a point O on one bank to a point P on the opposite bank, a distance b downstream. Happily for him, there is a rowing boat moored at O. If the stream flows with velocity u, what is the least speed with which the convict must row if he is to reach P? At this speed, in which direction must he row, and how long will it take him to arrive at P?

2.5. An equilateral triangular course ABC is marked out by buoys in a wide, straight river, with the buoy C being upstream and the line AB perpendicular to the stream. A motor boat follows a course which takes it from A back to A via B and C. If V is the speed of the boat in still water and u is the current speed, show that when the boat is moving along AB it points in a direction inclined at an angle $\sin^{-1}(u/V)$ to AB on the upstream side.

   What is the angle between BC and the direction in which the boat is pointed when it moves along BC? Show that when the boat turns at C it changes the direction in which it points by $120°$.

2.6. An aircraft which has a cruising speed of $480\,\mathrm{km\,h^{-1}}$ (480 kph) in still air takes off from Corfu to fly to London, which is 1200 km NW of Corfu. If the wind at cruising height is $120\,\mathrm{km\,h^{-1}}$ from the north, on what bearing must the aircraft fly and for how long will the flight last? If after the first hour of

---

* This is an umbrella with a straight handle.

the flight the wind speed drops to $30\,\mathrm{km\,h^{-1}}$, determine the position of the aircraft relative to London at the time it would have arrived had the wind speed not changed. Show that if the wind speed and direction remain fixed, all the destinations that can be reached by an aircraft travelling with constant air speed lie on a circle, the radius of which is independent of the direction of the wind.

2.7. A submerged submarine sights a hostile ship travelling due west. The submarine captain adjusts his course and speed to due NE and $30\,\mathrm{km\,h^{-1}}$ so that the ship bears NW of the submarine all the time. The submarine stops when it is $21\,\mathrm{km}$ from the ship so that it can fire a torpedo at $90\,\mathrm{km\,h^{-1}}$ in a direction due north at its target, the ship. What delay must the submarine captain allow between stopping and firing if he is to claim a sinking?

2.8. An aircraft leaves airfield A at 06:00 hrs to fly to airfield B, due north of A, at $400\,\mathrm{km\,h^{-1}}$. A second aircraft, with an average speed of $320\,\mathrm{km\,h^{-1}}$, is to leave airfield C, $480\,\mathrm{km}$ north and $320\,\mathrm{km}$ west of A, to intercept the first aircraft. Find the latest time of take-off for the second aircraft if it is to make the interception, and the direction in which it must travel.

Solutions

2.1. The maximum protection given by the umbrella will be provided when the umbrella* handle is aligned with the relative velocity. When the frame of reference (man and umbrella) is static, the umbrella handle is vertical, so that the rain is falling vertically. Taking the rain's velocity as $\mathbf{j}v$ and the walking speed as $\mathbf{i}5$, the angle of the umbrella handle to the vertical is given by $\tan\theta = 5/v$. With $\theta = 10°$, the value of v is $28.4\,\mathrm{km\,h^{-1}}$.

2.2. The frigate will be nearest to the battery when the line drawn to its course from the battery is normal to the course. Treating the frigate's course as the x-axis, we may write the velocities of the frigate and the cannonball as $5.7\mathbf{i}$ and $280\mathbf{j}$ respectively. The lay-off of the gunsight will be given by $\tan\theta = 5.7/280$, i.e. $\theta = 1.16°$, or $1°\ 10'$.

2.3. On his way to the platform, the velocity of the boy with respect to the river bank—his resultant velocity—is perpendicular to the bank, i.e. perpendicular to the direction of v in the velocity vector diagram. His swimming speed, u, forms the hypotenuse of a right angled triangle, the other sides of which are v and v_r, so that $v_r^2 = u^2 + v^2$. The return journey from the platform to the bank is represented once more by a right angled triangle in which $v_r^2 = u^2 - v^2$. The time that the boy spends in travelling from the shore and back is thus $t_1 = 100/[\sqrt{(u^2 - v^2)}]$. On the upstream and downstream legs of his refresher swim, his resultant velocities are $u - v$ and $u + v$ respectively, so that if the time for that immersion was t_2,

$$t_2 = [50/(u+v)] + [50/(u-v)] = (100u)/\left[\sqrt{(u^2 - v^2)}\right].$$

---

*Don't forget that this is an umbrella with a straight handle.

The ratio of these times is

$$t_2/t_1 = u/(u^2 + v^2).$$

If v were greater than u, the vector velocity diagram would not be right angled and v_r would have a downstream component. The boy could get to the platform by starting from a point on the bank upstream from it. He could not return to the same point on the shore after leaving the platform. During the refresher swim he would be swept downstream, whether he tried to swim up or down stream.

2.4. The boat's track must go from A to B so that it makes an angle $\theta = \tan^{-1}(b/a)$ with the normal to the shore. This is the direction of the boat's resultant velocity, and the minimum speed of rowing if the convict is to arrive at B occurs when the boat's velocity and the resultant velocity are directed perpendicular to each other. In that case we have $v_r = u \cos \theta = ua/[\sqrt{(a^2 + b^2)}]$, with the convict rowing in a direction perpendicular to AB. The resultant velocity of the boat is $u \sin \theta$ ($= ub/[\sqrt{(a^2 + b^2)}]$), and the length of its voyage is $l = \sqrt{(a^2 + b^2)}$. The time taken ($=$ distance/velocity) is $t = (a^2 + b^2)/ub$.

2.5. Take AB as the x-axis. Then, since we have the stream running perpendicular to AB, the boat's velocity is $V_r = iV_x + j(V_y - u)$, and the direction of V_r is given by $\tan \theta = (V_y - u)/V_x$. But V_r is parallel to AB and so $\theta = 0$, i.e. $V_y = u$. Since V_y and u are oppositely directed, the angle on the upstream side of AB is $\sin^{-1}(u/V)$.

For the leg BC of the circuit we may choose BC as our x-axis. BC is aligned at an angle of $30°$ to the bank, so that the component of u in the y-direction is $u \sin 30° = u/2$, with the result that $\sin \theta = (u/2V)$.

The two resultant velocities, just before and just after making the turn at C, are arranged at an angle of $60°$, so that the turn between the two tracks is $120°$. On the leg CA the boat again assumes an angle $\sin^{-1}(u/2V)$ with the track. This means that the change in the heading is the same as the turn between the tracks: $120°$.

2.6. The speed of an aircraft flying in still air is called its air speed (relative to the air) and is the same as the ground speed (measured relative to the ground) only when there is no wind. When the air moves as a wind the air speed and the ground speed are no longer the same, a situation which is entirely analogous to that of a boat travelling on a river, the velocity of which relative to the bank has to be compounded from the velocity of the boat and the velocity of the stream. The relation of the directions of the heading along which the aircraft is directed, its course, and the line which it follows on the ground—its track—are the same as the relation of the directions of the aircraft's air speed and the direction of its ground speed.

The aircraft travelling from Corfu to London must have its ground speed due NW, i.e. at N45°W, as shown in Fig. 2.8. The wind speed is directed from

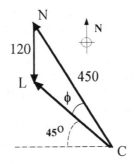

Fig. 2.8 The relation of ground speed, CL, air speed, CN, and the wind speed, NL, during the flight from Corfu to London.

the north so that the course of the aircraft has to be to the northward of its track. In Fig. 2.8 the angle CLN is 135°, so that the angle LCN is obtained from

$$\frac{480}{\sin 45°} = \frac{120}{\sin \theta},$$

leading to $\phi = 10.2°$. The course of the aircraft must be directed W55.2°N. Using the sine rule again, the ground speed is found to be $387.6 \, \mathrm{km \, h^{-1}}$, so that the time taken for the journey is 3 h 6 min.

When the speed of the wind changes, but its direction remains constant, there will be no alteration in the westward component of the ground speed; only in its northward component. In this case the northward component of the ground speed will increase by $90 \, \mathrm{km \, h^{-1}}$. The aircraft will travel with this new ground speed for 2.1 h, so that the aircraft will end up on the same longitude as London but some 189 km north of it, somewhere over Lincolnshire.

The question of the attainable destinations for the aircraft may be dealt with by assuming the wind speed to be directed along the y-axis: $u = |u| = u_y$. If the components of the constant air speed vector, V, along the x- and y-directions are taken as v_x and v_y, the components of the ground speed vector will have the form $v_{gx} = v_x$ and $v_{gy} = v_y + u$. The air speed vector has then the form $V = i v_{gx} + j(v_{gy} - u)$, so that the square of the magnitude of V is

$$V^2 = v_{gx}^2 + (v_{gy} - u)^2,$$

which defines a circle of radius V, centred at $v_{gy} = u$, in the velocity space the coordinates of which are (v_{gx}, v_{gy}). This circle is the location of the 'ends' of all the ground speed vectors which can be formed by combining an air speed, V, which is fixed in magnitude but may take any direction, and a wind speed, u. Since the allowed pairs of coordinates, (v_{gx}, v_{gy}), define a circle in their velocity space, the possible destinations reached during a given time of flight will also lie on a circle. This circle in position space for a flight time t will be described by the coordinates $x = v_{gx}t$, $y = v_{gy}t$. The radius, r, of this circle will be $r = Vt$, and its centre will be at $y = ut$.

2.7. The submarine's course is NE, and the ship is kept at a bearing of NW, so that the relative velocity and the submarine's velocity are arranged at right angles. The other angles in the right angled triangle in velocity space are each 45°, giving the ship's speed as $[30/(\cos 45°)] = 42.43$ km h^{-1}. When the submarine stops, the ship is 21 km NW of it, i.e. 14.85 km north and 14.85 km west of it. The velocity of the torpedo relative to the ship is 99.5 km h^{-1} directed N25.24°W, so that the torpedo must be fired when the ship passes through the point 14.85 km north and $(14.85 \tan 25.24°) = 7$ km W of the submarine. The time taken for the ship to reach this point is

$$[(14.85 - 7)/42.43] = 0.18 \, h = 11 \, \text{min} \, 6 \, s.$$

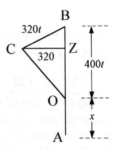

Fig. 2.9 The flight paths of the two aircraft in the 'interception limit'.

2.8. Let the time after take-off of the interceptor from C be t. A position vector diagram may be drawn as in Fig. 2.9. The shortest relative distance to be covered occurs when CO is perpendicular to the direction of flight C. This produces a right angled triangle with sides in the ratio 3:4:5, and so the angle between the flight directions is $\cos^{-1}(80t/100t) = 36.87°$. The distance OZ is given by $320 \tan 36.87° = 240$ km, which means that the distance of O from A is $(480 - 240) = 240$ km. The time of take-off of the interceptor from C is $(240/400)$ h after 06:00 hrs, i.e. 06:36 hrs. It's straightforward to show that the time of interception will be 08:16 hrs, although this wasn't requested.

3
Acceleration vectors

3.1 Acceleration vectors

So far, we have established that an object which is static may have its position represented by a vector. Even if the object is moving (as in Example 3, for instance) it may be located at a particular instant of time by a position vector, but then it requires a further vector, the velocity vector, to define its motion properly. In the static case the vector may be imagined as existing in normal, everyday, three-dimensional space. When the position vector changes with time we have introduced, in effect, a fourth dimension, time, which for most of us makes the visualization of the vector difficult, if not impossible. We avoided this difficulty by creating an imaginary three-dimensional velocity space (v-space) in which the extra variable, time, was subsumed into the space variables, so that the coordinate axes in the space were labelled with the components of the velocity in the three directions which were the usual x-, y-, and z-directions in space.

The next situation that we have to deal with is how to represent the case in which the velocity is changing. We have resolved the time dependence of position, movement in real space, by introducing velocity space to supplement our visualizations in real space. Now it seems not unreasonable to follow a path which is a direct analogue of the one we trod in creating velocity space from real space. This time, however, the changes with time take place in velocity space so that, say, in the time interval δt the velocity components describing the motion of the body change from v_x, v_y, and v_z to $v_x + \delta v_x$, $v_y + \delta v_y$, and $v_z + \delta v_z$. The slopes of the lines which define the velocity components in terms of time, dv_x/dt, dv_y/dt, and dv_z/dt, the first derivatives in time of the velocity components, may be used to define the axes in a new hypothetical space which, because the rate of change of velocity is acceleration, could be called an 'acceleration space'. The axes in this space are defined by $a_x = dv_x/dt$, $a_y = dv_y/dt$, and $a_z = dv_z/dt$, in just the same way that the derivatives of x, y, and z defined the axes in velocity space. An $a_x a_y$-plane is shown in Fig. 3.1, in which the direction of the acceleration vector $\boldsymbol{a}$ is given by $\theta = \tan^{-1}(a_y/a_x)$ and its magnitude is given by

$$a = |\boldsymbol{a}| = \left(a_x^2 + a_y^2 \right)^{1/2}.$$

This representation in acceleration space is not particularly helpful at this stage, but if we can say that a body is subjected to a constant acceleration $\boldsymbol{a}$ for a time δt, then the change in velocity which will be observed will be $\delta \boldsymbol{v} = \boldsymbol{a}\,\delta t$. For simplicity, take the case when the initial velocity of the body is zero. The change in velocity of the body in the time interval δt will be $\boldsymbol{a}\,\delta t$, and since this is a change from the state of zero velocity we can say that after the time interval δt the body will move with velocity $\boldsymbol{v} = \boldsymbol{a}\,\delta t$.

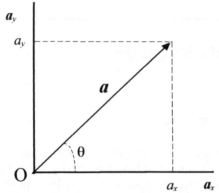

Fig. 3.1 An acceleration vector, *a*, in acceleration space.

If the initial velocity of the body is not zero, there is no reason why the velocity change $a\delta t$ should be in the same direction as *v*. In some cases the directions will be the same; for example, for a railway train on a straight track with gradients, the train is constrained to move along the line, so that accelerations and velocities must be collinear. On the other hand, a batsman striking a ball (bowled or pitched in his direction) towards the left or right boundaries of the field imposes on the ball an acceleration which is most certainly not in the direction of its velocity before it strikes the bat. The case in which velocity and acceleration are not collinear is illustrated in Fig. 3.2 in terms of a plane in velocity space, whence it can be seen that the velocity *v'* which characterizes a body after it has been subjected to an acceleration, *a*, non-collinear with *v* for a time δt, is simply the vector sum of *v* and δv $(= a\delta t)$, and is represented by the line joining the origin in *v*-space to the tip of the vector δv as drawn, so that

$$v'^2 = v^2 + (\delta v)^2 + 2v \cdot \delta v.$$

This means that in the case in which *v* and *a* are collinear—that is, when $\cos \theta = 1$

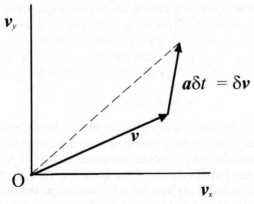

Fig. 3.2 The velocity space representation of a change in velocity which is not collinear with the initial velocity.

in the scalar product—we can drop the vector notation and take the square roots of both sides of the equation, to give

$$v' = v + a\,\delta t.$$

If a is constant this may be simplified further by substituting t for δt, to give

$$v' = v + at,$$

a relationship which is true only if v and a are collinear and if a is independent of time and position; that is, a is a constant.

Returning to Fig. 3.2, we notice that when a and v are not collinear, the velocity v' resulting from the change $v = a\,\delta t$ is not in the same direction as v. This means that a change in the direction of a velocity, even if its magnitude remains constant, requires an acceleration just as much as does a change in its magnitude. We might say, alternatively, that without an acceleration the velocity of a body will not change either in magnitude or direction: it will remain constant in both.

3.2 Centripetal acceleration and angular velocity

Example 9. A Formula One racing car is travelling along a straight at $200\,\text{km h}^{-1}$. At the end of the straight is a 30° bend, which is the arc of a circle of radius 300 m. What acceleration is required perpendicular to the path of the vehicle if it is to continue on its course round the bend without reducing speed?

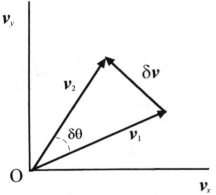

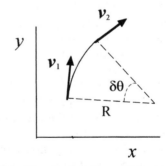

Fig. 3.3 The velocity vector diagram to find the velocity change occurring when a velocity has its direction altered from v_1 to v_2 through an angle $\delta\theta$.

Fig. 3.4 The track of a vehicle travelling with constant speed along a circular arc of radius R.

The first step in this problem is to draw a velocity diagram in velocity space, as in Fig. 3.3, to find a suitable expression for the change in velocity which takes place. Let the velocity v_1 be altered in direction by a small angle $\delta\theta$, while its magnitude remains fixed. The triangle formed by v_1, v_2, and δv is an isosceles triangle and the length of its short side is given by

$$\delta v = 2v \sin(\delta\theta/2).$$

For small angles such as $\delta\theta$, $\sin(\delta\theta/2) \approx \delta\theta/2$, giving

$$\delta v = v\delta\theta.$$

If the acceleration producing the velocity change is a, then $\delta v = a\delta t$, so that

$$a = \delta v/\delta t = v(\delta\theta/\delta t).$$

In position space (*xyz*-space) we may draw the path of the vehicle in the time it has taken to alter its direction of motion by $\delta\theta$ as shown in Fig. 3.4, in which two velocity vectors (not the same ones as in Fig. 3.3) are sketched to indicate the relation of their orientation change and the position of the vehicle. The length of the arc representing the path of the vehicle may be expressed as $R\delta\theta$ and as $v\delta t$ respectively. Since these two must be the same,

$$\delta\theta/\delta t = v/R$$

so that, on substituting back into the expression for the acceleration, we find

$$a = v^2/R,$$

which is the acceleration required to produce the change in velocity consequent on travelling with constant speed v around an arc of radius R.

The racing car has a velocity of $200\,\text{km}\,\text{h}^{-1}$, or $55.55\,\text{m}\,\text{s}^{-1}$. The radius of the curve is $300\,\text{m}$, with the result that $a = 55.55^2/300 = 10.3\,\text{m}\,\text{s}^{-2}$. For comparison, the '0–60 mph in 6 s' so admired by purchasers of 'performance' cars corresponds to a uniform acceleration of about $5\,\text{m}\,\text{s}^{-2}$.

We have eliminated the quantity $\delta\theta/\delta t$ from this example without taking its limit as $\delta t \to 0$. This is appropriate provided that the speed, v, and the radius, R, of the arc remain constant with time, but would be inadequate if either of them changed. Then we would have to take the limit as $\delta t \to 0$, given the equality $d\theta/dt = v/R$. This again leads to $a = v^2/R$. The importance of this second case, that of the infinitesimal limit, is that as $\delta\theta$ tends towards zero the other angles in the velocity triangle approach $90°$ more and more closely. In the limit of small δt this angle *is* $90°$, so that the acceleration is directed perpendicularly to the velocity, towards the centre of the arc, as may be seen by amending Fig. 3.4 suitably. This is why 'the acceleration perpendicular to the path' was required in the example. This acceleration, required to keep a body moving along an arc with constant speed, is often called the centripetal (centre-seeking) acceleration because it is directed 'towards the centre'.

The occurrence of the relation $v = R(d\theta/dt)$ suggests that we might well define the rate of change of angular position, $d\theta/dt$, as the angular velocity, ω; that is, $\omega = d\theta/dt$. Such a definition would give us the choice, in talking about movement along curved lines, of discussion in terms of the linear velocity, $v = dx/dt$, or in terms of the angular velocity $\omega = d\theta/dt$. Since the movement is along a curve, the use of ω is often more convenient. In that case an acceleration $R\omega^2$ is required to maintain an angular velocity ω round a curve of radius R. This acceleration is, of course, centripetal. When we define arc length in terms of radius and angle as $R\theta$, the angle θ is measured in terms of radians (for example, the circumference of a circle is $2\pi R = R \times 2\pi$). Angular velocity is measured, correspondingly, in radians s^{-1} or, since radians are dimensionless, simply as s^{-1}.

3.3 Position–velocity–time relations

Uniform linear motion in the one-dimensional case is described in terms of time, t, position, x, and velocity, v. We have introduced acceleration, a, to enable us to deal with changing velocities. Because $v = dx/dt$ we may extend our definition of the acceleration to

$$a = \frac{dv}{dt} = \frac{d(dx/dt)}{dt} = \frac{d^2x}{dt^2},$$

where the right-hand side of the equation is the derivative with respect to time of the derivative of x with respect to time, the second derivative of x with respect to t. As a simple example, if $x = t^2$, then $dx/dt = 2t$, and $d^2x/dt^2 = 2$; the second derivative is obtained by applying the operation of differentiation twice.

There is a second way of looking at a which produces a different expression for the acceleration. This is obtained by taking dv/dt to be the product of two derivatives, one of v with respect to x and the other of x with respect to t, thus

$$a = \frac{dv}{dt} = \frac{dv}{dx}\frac{dx}{dt} = v\frac{dv}{dx},$$

so that the acceleration may be obtained by multiplying the velocity and the rate of change of velocity with position.

Just as where we had a velocity we could subject it to a process of integration (with suitable boundary conditions) to give displacement as a function of time, so in the case of acceleration the acceleration may be integrated to give the velocity as a function of time or as a function of position. Integration of a with respect to t gives

$$\int_{t_1}^{t_2} a\,dt = \int_{t_1}^{t_2} \frac{dv}{dt}\,dt = \int_{v_1}^{v_2} dv,$$

where v_1 and v_2 are the velocities which correspond to the times t_1 and t_2, so that when a is constant

$$v_2 - v_1 = a(t_2 - t_1).$$

If we choose $t_1 = 0$ and say that the initial velocity is v_1, then

$$v_2 = v_1 + at,$$

which may be integrated with respect to time to give

$$x = v_1 t + \tfrac{1}{2}at^2.$$

Alternatively, we could integrate our x-dependent expression for the acceleration

$$\int_{x_1}^{x_2} a\,dx = \int_{x_1}^{x_2} v\frac{dv}{dx}\,dx = \int_{v_1}^{v_2} v\,dv,$$

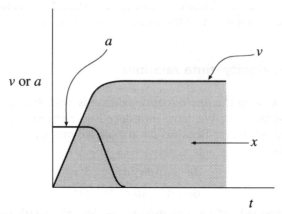

Fig. 3.5 A graphical representation of acceleration, $a = (dv/dt$, velocity, $v = dx/dt = \int a\,dt$, and of distance, $x = \int v\,dt$ (the shaded area under the v curve), all as functions of time, t. In this graph the initial (boundary) conditions are that $a = $ constant, $v = 0$, and $x = 0$, when $t = 0$.

where v_1 and v_2 are the velocities at the positions x_1 and x_2 respectively, giving when a is constant

$$a(x_2 - x_1) = \tfrac{1}{2}(v_1^2 - v_2^2).$$

This time, taking $x_1 = 0$ so that the initial velocity is v_1, we have

$$v_2^2 = v_1^2 + 2ax.$$

Acceleration is the gradient of the velocity–time curve and velocity is the gradient of the position–time curve. Equally, velocity can be represented as $v = \int a\,dt$, so that v is the area under the acceleration–time curve. Similarly, $x = \int v\,dt$ is the area under the velocity–time curve. These relations are illustrated in Fig. 3.5.

Example 10. A train approaching a station is slowing down with uniform accelera-tion (or is retarded uniformly). If it is seen to take 40 s to pass along an 800 m stretch of track, and 60 s to travel through the next 800 m, how far away is it from the station at which it will come to a halt?

The information gives distance in terms of time, so that the equation to use is

$$x = v_1 t + \tfrac{1}{2}at^2,$$

in which a and v_1 are unknown. Since two data sets are available, we may eliminate either of these unknown quantities and obtain the other. At the end of the first 800 m

$$800 = (v_1 \times 40) + (\tfrac{1}{2}a \times 1600).$$

The train takes 100 s to cover 1600 m (adding together the two times and the two distances so that v_1 is the same for both equations); that is,

$$1600 = (v_1 \times 100) + (\tfrac{1}{2}a \times 10\,000).$$

Multiplying the first of these equations by 2.5 and subtracting one from the other to eliminate v_1 gives

$$-400 = \tfrac{1}{2}a \times 6000 \quad \text{or} \quad a = -2/15 \, \text{m s}^{-2},$$

so that

$$v_1 = 23.67 \, \text{m s}^{-1}.$$

Then, using

$$v_2^2 = v_1^2 + 2ax,$$

we find that

$$x = 2.1 \, \text{km}$$

from the initial position of the train, or 500 m from the end of the second 800 m stretch of line. It will arrive at the station approximately three minutes after it passes the initial position.

Example 11. The crew of a police car waiting in a lay-by observe a car exceeding the 80 kph (1 kph = 1 km h^{-1}) speed limit by 32 kph. As it passes them they set off in pursuit, accelerating at 6.5 m s^{-2} for 6 s and then maintaining a uniform speed. How far will they travel before they overtake the speed merchant?

The acceleration of the police car takes it to a speed of $v = (6.5 \times 6) = 30 \, \text{m s}^{-1}$, and it has then travelled a distance of $x = \tfrac{1}{2}(6.5 \times 36) = 117 \, \text{m}$. In those 6 s the speeding motorist has travelled a distance

$$v_m t = [(112 \times 1000)/3600] \times 6 = 186.67 \, \text{m},$$

so that the distance then separating the two cars is 69.67 m. The relative velocity of the vehicles is 7.8 m s^{-1}, and the additional time taken to overtake is 8.8 s. In this time, the police car will have covered a further 343.6 m. In all, it will have travelled 466.6 m before it overtakes the offending motorist, roughly 15 s after starting.

Example 12. The reaction time of an alert car driver, the time it takes him to notice an event and apply the brakes, accelerate, or change direction, is about half a second. If such a driver, travelling at 48 kph, approaching an intersection with a set of lights, sees the light turn red, how far will he travel before stopping if his application of the brakes causes a deceleration of 2 m s^{-2}?

During the driver's reaction time, the car travels a distance

$$vt = (48\,000/3600) \times 0.5 = 20/3 \, \text{m}.$$

The distance travelled during the deceleration is obtained from $v_2^2 = v_1^2 + 2ax$; that is,

$$0 = (20/3)^2 - 2.2x,$$

giving $x = 44.4 \, \text{m}$.

The total stopping distance is thus 51.07 m. The time taken before stopping would be nearly 7.2 s, so that the driver has made a leisurely stop. If he had been watching the light sequence he might have accelerated to pass safely through the intersection, or made an even more leisurely stop.

In just the same way that we have relative velocity of two bodies, so two accelerating bodies have a relative acceleration which is the vector sum of the two accelerations, or the sum and difference of their magnitudes if they are travelling along the same line. The ability to transfer between different frames of reference by addition of displacements, of velocities, or of accelerations is restricted to cases in which the velocities concerned are much smaller than the velocity of light. Going from one frame of reference to another in this case is called a Galilean transformation; this is distinct from the relativistic case, in which a Lorentz transformation must be used. Here we will be concerned only with Galilean transformations.

Exercises

3.1. A laboratory testing machine used to examine the effects of accelerations on the human body consists of a horizontal arm, at the end of which is attached a seat in which a volunteer is seated. The arm is rotated in a horizontal plane about a vertical axis which is 5 m from the seat. What rate of rotation will subject the 'guinea pig' to an acceleration of $100\,\mathrm{m\,s^{-2}}$?

3.2. An aircraft landing on a horizontal runway 3 km long has a velocity of 250 kph as it touches down. The traffic controller tells the pilot to leave the runway along a taxiing track which is situated half-way along the runway. The speed of the aircraft as it turns off the runway must not be greater than 60 kph. What is the minimum deceleration of the aircraft after landing if the pilot is to conform to the controller's instruction?

3.3. A stage of the Tour de France ends on a straight, level road leading into the centre of a town. When the race has 500 m to run, the leader of the breakaway group, travelling at 48 kph, is passed by a rival who has increased his speed to 56 kph. Provided that the first cyclist reacts instantly, at what rate must he accelerate in order to ensure that he wins the stage?

3.4. A pit cage in a coal mine performs the first part of its descent with a uniform acceleration $a\,\mathrm{m\,s^{-2}}$ and the remainder of its descent with a uniform retardation of $2a\,\mathrm{m\,s^{-2}}$. If the working level is 1 km below ground level and the time taken in the descent is 25 s, what is the value of a?

3.5. A high speed lift carries tourists to the top of a 675 m high observation tower in 45 s. For a distance of the first quarter of its ascent, the lift is accelerated uniformly and during the last quarter distance of its ascent it is retarded uniformly, the acceleration and deceleration having the same magnitude. With what speed is the lift travelling when it passes the central point of the tower?

3.6. A driver wishes to travel from one motorway service station to another, which is at a distance x from the first. He makes his journey in a time t by first of all accelerating with constant acceleration, a, then by travelling with constant velocity, and finally by decelerating with constant retardation, a. Show that the time for which he travels with constant velocity is $\sqrt{[t^2 - (4x/a)]}$.

3.7. A passenger in a train alleviates his boredom by noting the times at which he passes successive telegraph poles erected at the side of the line. He knows they are spaced at 50 m intervals, and he observes that the train has accelerated because the time intervals between successive pairs of poles are 3 s and 2.7 s. What is the acceleration of the train, and at what speed is it travelling after this brief spasm of acceleration?

3.8. The maximum speed attainable by a sprinter is (approximately) $10.5\,\mathrm{m\,s^{-1}}$. If he can sustain a maximum acceleration of $1\,\mathrm{m\,s^{-2}}$ during a 100 m dash, at what speed must he leave his starting blocks if he is to break 10 s?

Solutions

3.1. The acceleration of $100\,\mathrm{m\,s^{-2}}$ is produced by the arm whirling around its support. In terms of angular velocity, $a = R\omega^2$, $\omega^2 = 100/5$ and thus $\omega = 4.47\,\mathrm{s^{-1}}$. This converts to revolutions of the arm by division by 2π so that the arm makes 0.712 rps, or 42.7 rpm.

3.2. $1\,\mathrm{m\,s^{-1}} = 3.6\,\mathrm{km\,h^{-1}}$, so that, using the equation $v_2^2 = v_1^2 - 2ax$, we obtain $(69.44^2 - 16.67^2) = (2 \times 1500a)$, leading to $a = 1.52\,\mathrm{m\,s^{-2}}$.

3.3. The overtaker maintaining his speed will cross the finishing line in $(500/15.56) = 32.1\,\mathrm{s}$. If the overtaken cyclist is to get to the line in the same time, then, for him,

$$500 = (13.33 \times 32.1) + [0.5 \times a \times (32.1)^2], \quad \text{i.e.} \quad a = 0.14\,\mathrm{m\,s^{-2}}.$$

If the overtaken cyclist is to win the stage he must accelerate at more than $0.14\,\mathrm{m\,s^{-2}}$ and his speed at the finishing line will be more than 64.2 kph.

3.4. The depth of the shaft is $h = h_1 + h_2$, where h_1 and h_2 are the distances covered during acceleration and deceleration respectively. The times occupied by acceleration and retardation are t_1 and t_2, so that the total time taken to reach the working level is $t = t_1 + t_2$. The initial and final velocities are zero, so that

$$h_1 = 0.5 \times at_1^2 \quad \text{and} \quad h_2 = vt_1 - (0.5 \times 2at_2^2),$$

where v is the maximum velocity attained by the cage. Since $v = at_1$,

$$h_2 = at_1t_2 - at_2^2.$$

Because $v - 2at_2 = 0$, $t_1 = 2t_2$, so that $h_2 = a2t_2^2 - at_2^2 = at_2^2$ and $h_1 = \frac{1}{2}at_1^2 = 2at_2^2$. This leads to $h - h_1 + h_2 = 3at_2^2$, but since $t_2 = t/3$ we have $h = (at^2)/3$, giving $a = 4.8\,\mathrm{m\,s^{-2}}$. The maximum velocity attained is $40\,\mathrm{m\,s^{-1}}$, i.e. 144 kph.

3.5. In this case we have three vertical distances, $y_1 = y_3 = \frac{1}{4}h$, and $y_2 = \frac{1}{2}h$, with the corresponding times t_1, t_2, and t_3. The uniform velocity $v = at_1$. Then

$$y_1 = 0.5 \times at_1^2 = 0.5 \times t_1^2(v^2/2y_1) \quad \text{or} \quad y_1 = \frac{1}{2}vt_1.$$

We have $y_2 = vt_2$, so $y_1/y_2 = t_1/2t_2$. Since $2y_1 = y_2$, we have $t_1 = t_2$. Furthermore, $y_3 = y_1$, so that $\frac{1}{2}at_1^2 = (vt_3 - \frac{1}{2}at_3^2)$ or $a(t_1^2 - 2t_1t_3 + t_3^2) = 0$, i.e. $t_1 = t_3$, with the result that $t_1 = t_2 = t_3 = 15$ s. Substitution for t_1 and y_1 in the equation $y_1 = \frac{1}{2}at_1^2$ leads to $a = 1.5 \, \text{m s}^{-2}$ and $v = 22.5 \, \text{m s}^{-1}$.

3.6. The time taken in the acceleration and retardation must be the same, since they both have the same magnitude, a. The total time taken in the journey is $t = 2t_1 + t_2$. The distances x_1 and x_2 will be the same as well: $x_2 = x_1 = \frac{1}{2}at_1^2$. The distance travelled at constant speed is $vt_2 = at_1t_2$, and so

$$x = at_1^2 + at_1t_2 \quad \text{or} \quad x/a = t_1^2 + t_1t_2 = 0.25(t - t_2)^2 + 0.5(t - t_2)t_2.$$

Therefore $4x/a = t^2 - t_2^2$, giving $t_2 = \sqrt{[t^2 - (4x/a)]}$.

3.7. For the first interval,

$$50 = 3v_1 + (0.5 \times a \times 9).$$

For both intervals,

$$100 = 5.7v_1 + [0.5 \times a \times (5.7)^2].$$

The first equation can be modified to $90 = 5.7v_1 + 8.55a$, so that $a = 0.65 \, \text{m s}^{-2}$, and finding v_1 from one of the equations and then using

$$v_2 = v_1 + (0.65 \times 5.7)$$

leads to $v_2 = 69.8 \, \text{kph}$.

3.8. The distance, x, travelled in time t is $x = v_1t + 0.5at^2$. With $a = 1 \, \text{m s}^{-2}$, the maximum value of the second term is 50, while the value of x is 100. $10v_1 = 50$, this gives a starting speed of $5 \, \text{m s}^{-1}$. The sprinter's finishing speed would be $v_2 = v_1 + at = 15 \, \text{m s}^{-1}$. Since the maximum speed with which a man can run is about $10.5 \, \text{m s}^{-1}$ (32 kph, or 24 mph), this is not a realistic problem.

If the sprinter started from a standing position his initial velocity would be zero, and if he completed the 100 m in Ts then $100 = [0.5at^2 + v(T - t)]$, where t is the acceleration time. From a standing start we might expect a range of T from 11 to 12.5 s. Taking $a = v/t$ and v as 10.5 leads to the following table of standing start 100 m times, accelerations, acceleration times, and acceleration distances:

T (s)	$a \, (\text{m s}^{-2})$	t_1 (s)	x_1 (m)
11	3.5	3	15.75
11.5	2.66	3.95	20.7
12	2.12	4.95	26.0
12.5	1.76	5.95	31.2

These figures suggest that the maximum acceleration that a sprinter can maintain is probably about $3 \, \text{m s}^{-2}$. Assume that this is the case. Then, with a starting speed of $5 \, \text{m s}^{-1}$, the uniform acceleration of $3 \, \text{m s}^{-2}$ occurs during the first 1.83 s of the race. The rest, run at a constant speed of $10.5 \, \text{m s}^{-1}$, takes 8.17 s. A maximum speed of $10.6 \, \text{m s}^{-1}$ with the same starting speed and acceleration would give an advantage of nearly 0.8 m at the tape.

4
Centres of mass

4.1 Inertial mass

So far, we have examined systems (automobiles, aircraft, ships, etc.) in terms of their positions, their velocities, and the accelerations to which they are subjected. We have developed the subgroup of mechanics known as kinematics. Kinematics is a phenomenology which bypasses the question fundamental to mechanics, 'What is the origin of acceleration?' Since our avowed topic is mechanics, we need a new beginning to provide an answer to that question.

A good starting point for the study of mechanics—just as in most topics in physics—is to look at the results of a simple experiment. As the subject of the experiment, take a light rigid rod and attach to its ends two equally sized spheres made from the same material, say from polyethylene. Then place the rod on a knife-edge or suspend it with a loop of string. Unless the point of contact between the rod and the knife-edge (or the point of suspension from the string) is exactly at the centre of the rod, the rod will rotate so that the sphere further away from the point of support moves downwards. Only when the point of suspension is at the centre of the rod will it be balanced so that neither of the spheres moves. After the rod has been balanced on its support, the next step in the experiment is to remove one of the polyethylene spheres and replace it with an iron sphere of equal dimensions; that is, of the same radius. All attempts to balance the system with the point of suspension at the centre of the rod will then fail.

In either part of this experiment the geometrical symmetry of the rod–sphere system has not been altered. All its components have the same dimensions in both parts, and yet the results of observing the behaviour of the system after changing the material of one of the spheres from polyethylene to iron are quite different. The only change is the substitution of a heavier iron sphere for one of the polyethylene ones. The lack of balance, manifested as a rotation of the rod about its point of support, must be related to the difference in heaviness between the two spheres, one at each end of the rod. We express this difference by saying that the rotation occurs because the two spheres have different inertial masses, and because the iron sphere is heavier its inertial mass is the larger of the two masses (for convenience the inertial mass is referred to usually as the mass—the word inertial is left as implicit). The unit of mass is the kilogram, (kg).

When the iron sphere is substituted for one of the polyethylene ones it is found that the rod can be balanced about a point of suspension, but this is no longer at the centre of the rod. This new state of balance is obtained when the point of suspension is very much closer on the rod to the iron sphere than it is to the polyethylene one. At the point of balance no movement occurs so that the two masses, one at each end of the rod, must have equal and opposite effects. The position of the point of balance on the rod is called the centre of mass of the rod–sphere combination. When the spheres are made of identical material, the point of balance lies exactly half-way between them. When they are made of different materials, the balance point is much closer to the end of the rod at which the heavier sphere is situated. When the system is balanced in this way the rod is static—it is not moving—so that this state is described as a state of static equilibrium. This state will be maintained unless there is some external reason for its being disturbed.

Let's now complicate our experiment slightly by attaching the two polyethylene spheres to the rod at equal distances from its centre but not at its ends, and again balance the rod on its support. Then enclose the system in an opaque box, with only the ends of the support protruding through slots in its ends. Given that the mass of the enclosed system is twice the mass of one of the spheres, what can we deduce about the distribution of mass in the rod–sphere system, or about the position of the system's centre of mass? All that we can say with certainty is that the system has a mass $2m$, where m is the mass of one of the spheres, that the support is situated at the centre of mass of the system, and that the support is supporting the mass $2m$. We are unable to tell even whether the system consists of the two spheres on the rod balanced on the support. It could be that there is simply one sphere of mass $2m$ which, because there is no other sphere to balance it, must be situated on the rod at the position of the support. It might be that there are several spheres, the combined mass of which is $2m$, attached at various points along the rod and the support is situated at the centre of mass of this multi-sphere system. In all possible equilibrium cases the support is carrying the mass $2m$ and is situated at the centre of mass of the rod–sphere system.

4.2 Discrete mass systems

When we are dealing with a two mass system (for example, the rod–sphere combination) it's convenient to choose a coordinate axis, say x-axis, along the line joining them. The origin of the axis ($x = 0$) may be chosen to taste; at a point between the two masses, coincident with one of them, or even to one side or the other of the mass–rod combination. When this system is in the equilibrium state (it is balanced) the whole effect of the two masses, say ($m_1 + m_2$), is to act through the support which is at a position X in the one-dimensional coordinate system. Put in another way, the rod–sphere system is behaving as if its mass was a single mass of mass $m_1 + m_2$ positioned at the support. In this case we make the statement that the position of the centre of mass of the system is given by

$$m_1 x_1 + m_2 x_2 = (m_1 + m_2)X,$$

where x_1 and x_2 are the position coordinates of the masses $m_1 + m_2$. This is obviously correct in the case in which the spheres have equal mass, for then $m_1 = m_2$ and

$$X = [(x_1 + x_2)/2];$$

in other words, the centre of mass is half-way between the two masses, as observed above. The definition of the position of the centre of mass also predicts that when the two spheres are of different materials the centre of mass will be closer to the heavier of the two spheres, consistent with the observation made on the asymmetric (polyethylene/iron) system of masses. As an illustration of this point, choose $x_1 = 0$, $x_2 = x$, and the centre of mass at X, then, for example, if $m_1 = 7m_2$ we have $X = (x/8)$. The centre of mass is one eighth of the way along the rod, much closer to the heavier of the two masses then to the light one.

We can extend our two-body definition of the position of the centre of mass to the case of many bodies by saying that the ith body $(i = 1, 2, \ldots, n)$ has a mass m_i and is situated at the position x_i so that

$$m_1 x_1 + m_2 x_2 + \cdots + m_n x_n = (m_1 + m_2 + \cdots + m_n)X$$

or

$$X = \frac{\Sigma m_i x_i}{\Sigma m_i},$$

where the summations are taken over the values of i from 1 to n.

Example 13. Particles of mass 2, 3, 6, and 9 kg are placed along a straight line at distances of 1, 2, 3, and 4 m from a point A, also lying on the line. What is the distance from A of the centre of mass of the system?

This problem is an illustration of the extension of the idea of centre of mass of a two-body system to the case involving more than two bodies. The relation giving the position of the centre of mass now becomes

$$X = \frac{m_1 x_1 + m_2 x_2 + m_3 x_3 + m_4 x_4}{m_1 + m_2 + m_3 + m_4},$$

which is a particular case of the general form

$$X = (\Sigma m_i x_i)/(\Sigma m_i),$$

where the summation sign Σ indicates that the quantities $m_i x_i$ and m_i are added over all the values of the suffix i. Substitution of the numbers into the formula gives

$$X = 3.1 \, \text{m}.$$

Example 14. Masses of 1, 2, and 3 kg are placed at the corners of an equilateral triangle of side 9 m. What is the position of the centre of mass relative to the position of the 1 kg mass?

This problem represents an extension of the idea of centre of mass into two dimensions. We may draw the triangle (shown in Fig. 4.1) so that the larger masses lie on the y-axis while the 1 kg mass lies on the y-axis. The angles in the triangle are all the same, 60°, so that the coordinates of the masses are $(7.8, 0)$, $(0, 4.5)$, and $(0, 4.5)$ in order of increasing mass. (What is the length of the unit vector in this coordinate system?) The x-coordinate of the centre of mass is given by

$$X = \frac{(7.8 \times 1) + (0 \times 5)}{6} = 1.3 \, \text{m}.$$

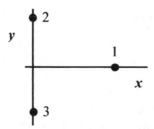

Fig. 4.1 The coordinate system for three masses at the vertices of an equilateral triangle.

Because the masses positioned along the *y*-axis are not equal, the centre of mass cannot be situated on the *x*-axis. It is necessary to find the *y*-coordinate of the centre of mass from

$$Y = \frac{(2 \times 4.5) + (3 \times (-4.5)) + (0 \times 1)}{6} = -0.75 \, \text{m}.$$

Within the chosen coordinate system the centre of mass has the coordinates $(1.3, -0.75)$, while the 1 kg mass is situated at the point $(7.8, 0)$. The distance of the centre of mass from the 1 kg mass is then

$$d = (7.8 - 1.3)^2 + (0 - (-0.75))^2 = 6.54 \, \text{m}.$$

Equally as well, we could have specified the positions of the masses by position vectors so that

$$R = \frac{m_1 r_1 + m_2 r_2 + m_3 r_3}{m_1 + m_2 + m_3}.$$

It might appear that choosing the position of the smallest mass to be at the origin and substituting the magnitudes of the masses and the vectors would give the distance from the centre of mass to the 1 kg mass directly. This is not the case: the vectors have to be written in component form in order to perform the calculation. Remember that the vector form of R would be written as

$$R = X + Y = iX + jY,$$

so that we would have to deal with the components of the r's and we would have finished up with the calculations that we made above.

4.3 Laminas

Example 15. A circular disc 180 cm in diameter has a circular hole of 60 cm diameter cut out of it. If the centre of the hole is 40 cm from the centre of the disc, by what distance has the centre of mass of the disc moved?

The first observation to be made here is that a symmetrical system of masses will have its centre of mass at its centre of symmetry. This means that we may assume that the centre of mass of the circle (uncut), which is a system of masses arranged symmetrically about its centre, lies at the centre of the circle. For convenience,

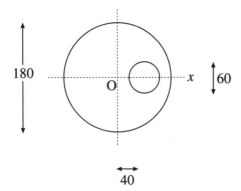

180

O

x   60

40

Fig. 4.2   The circle with a circular hole cut from it.

designate this centre of the circle as the origin $(0,0)$, and the x-axis as the line joining the centre of the circle to the centre of the hole, as shown in Fig. 4.2. The same argument may be used to establish that the centre of mass of the circle of material which was cut from it was situated at the centre of the circular hole which resulted from the cutting; in other words, at the position $(40, 0)$. The masses of the disc (m_d) and of the material removed to form the hole (m_h) will be proportional to the areas of the disc and the hole; that is, to the squares of their radiuses, $(3/1)^2 = 9$. The centre of mass of the disc–hole combination is situated along the x-axis because the x-axis is a line of symmetry (it was chosen with this simplification in mind). The x-coordinate of the centre of mass is obtained from

$$X = \frac{m_d x_d + (-m_h) x_h}{m_d + m_h}$$

$$= [(-1/9) x_h]/(8/9) = -(40/8) = -5 \, \text{cm}.$$

The centre of mass is situated 5 cm from the centre of the disc along the diameter common to the disc and the hole, in the half of the disc opposite to the hole.

This example is of interest because it illustrates that we have extended the use of the formula for finding the position of the centre of mass. We have made the implicit assumption that the mass of the disc and the mass of the material taken away from it to form the hole may be considered, in effect, to be acting at their respective centres of mass. Then we have dealt with the problem as if it were a simple two-body problem, although in this case the 'mass' of the hole has to be taken as negative because it was taken away from the disc. This is a demonstration of the principle that the centre of mass of a system of two or more components which are not point masses may be calculated by assuming that each component behaves as if its mass were a point mass positioned at its centre of mass.

Example 16.   A flat plate is cut in the form of a triangle ABC (another description of this is 'a triangular lamina ABC') with angles A, B, and C, the values of which are restricted only by the condition $A + B + C = 180°$. What is the position of the centre of mass of the triangle?

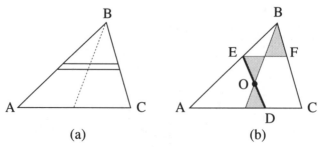

Fig. 4.3 (a) The construction needed to find the position of the centre of mass of the triangle ABC. (b) The second construction to locate the position of the centre of mass of the triangle ABC.

The triangle is shown in Fig. 4.3(a), with AC horizontal. We may divide the triangle up into strips parallel to AC, as shown in Fig. 4.3(a), and make the statement that the centre of mass of each strip is situated at its centre. The centre of mass of the triangle will lie on the line joining B to the centre of AC, drawn dashed in the diagram. We might have drawn the strips parallel to AB or to BC and seen that the centre of mass of the triangle lies on the line bisecting AB and joined to the opposite angle C, and along the line bisecting BC and joined to the opposite vertex, A. The centre of mass must be situated at the point at which these three bisecting lines meet with one another. If we can find the position of this point in the triangle we will have solved our problem. This is a geometrical exercise which begins with our redrawing the triangle as shown in Fig. 4.3(b) and marking the point O at which the bisecting lines intersect. Then we draw a line DE parallel to BC and passing through O and a line EF passing through E and parallel to AC. This procedure leads to the formation of the three triangles shown shaded in Fig. 4.3(b). OD and OE have the same length because AO is one of the bisecting lines; the two lower shaded triangles have the same values for their three angles so that they are congruent. By a similar argument, since OB bisects EF, the two upper triangles are congruent too. This means that the three shaded triangles are all congruent, having the same shapes and sizes. The centre of mass of the triangle is thus situated at a distance $l/3$ along the bisector of angle B from AC, where l is the length from AC to B of that bisector. Since each of the congruent triangles has the same vertical height, the result could be expressed as well in terms of the vertical height of ABC. The centre of mass of the triangle ABC is situated at a point which is one-third of the vertical height of the triangle measured from the base of the triangle. This result is true whichever of the triangle sides—AB, BC, or CA—is chosen as base, so that the two parameters required to locate the centre of mass may be found quite simply.

This problem demonstrated that geometrical arguments, beyond the centre of symmetry statement made previously, may be used to find the position of a centre of mass. Although the centre of mass of a triangle may be located quite simply if the triangle is right angled or shows symmetry, an analytical solution for an arbitrary triangle would be relatively complicated. The idea mooted in Example 15, that of dividing a continuous body into small elements and treating them as individual masses, has been made explicit by using strip mass elements.

4.4 Continuous bodies

We have already dealt with two cases of laminas in which the mass distribution is no longer discrete but continuous. These were resolved using particular methods, one of treating the lamina as if it were two discrete masses, and the other of using the geometrical properties of the lamina. Such methods are often not convenient to use, and we have to provide an answer to the question as to how we can deal with continuous bodies such as laminas or three-dimensional solids in general. To begin with, we define the position of the axes which we will use for reference. Once this has been done the continuous body is imagined to be composed of pseudo-discrete mass elements δm situated at the point (x, y, z) [or (r, θ, ϕ)], as illustrated for a two-dimensional case in Fig. 4.4. If these mass elements were truly discrete the relation defining, for example, the x-coordinate of the centre of mass would have the form $X = (\Sigma \, \delta m_i x_i)/(\Sigma \, \delta m_i)$, where i is the subscript labelling the ith mass element. Because the mass elements are not discrete, the summation process is no longer appropriate. In place of the summations, we need to use integration processes and the position of the centre of mass (along the x-direction) takes the form of a ratio of two integrals, $X = [(\int x \, dm)/(\int dm)]$. The integrals are evaluated by expressing dm in terms of the x-, y-, and z-variables and the density of the material from which the lamina or body is composed.

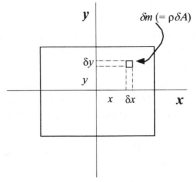

Fig. 4.4 A diagram showing one of the mass elements, δm, at the position (x, y) for a rectangular lamina. This is a two-dimensional case of the method used for locating the centre of mass of a continuous body, because the lamina lies in the plane $z = 0$. In this illustration, δm is the product of the area density, ρ, of the material of the lamina, and the area $\delta A \, (= \delta x \delta y)$ of the element.

Example 17. A piece of uniform thin wire is bent into the shape of a circular arc so that its length subtends an angle 2α at the centre of the circle which is of radius r. Locate the centre of mass of the bent wire.

The bent wire is illustrated in Fig. 4.5(a). Because of its symmetry, the centre of mass must lie along the line which joins its centre to the centre of the circle. Take this line as the x-axis and choose the origin at the centre of the circle. The problem is one of identifying the x-coordinate of the centre of mass, since we have seen that $Y = 0$. It is resolved by dividing the wire into elements of equal mass (meaning that

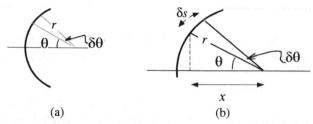

Fig. 4.5 (a) The uniform wire bent into the arc of a circle. (b) The definition of δs, $\delta \theta$, and x for the bent wire.

the elements have equal lengths because the wire is uniform). Let's say that each of these elements has a mass δm (as shown in Fig. 4.5(b)), implying that the elements are very small indeed. Because the elements are all of the same length we may say that each has a length δs, which in terms of the angle, θ, subtended at the centre of the circle may be written $r \delta \theta$.

$$\delta s = r\, \delta \theta.$$

The x-coordinate shown in Fig. 4.5(b) for the element δm is simply $r \cos \theta$, so that the coordinate of the centre of mass would be given by the sum

$$X = \frac{\Sigma x \delta m}{\Sigma\, \delta m} = \frac{\Sigma x \delta s}{\Sigma\, \delta s}$$

if the mass elements were discrete. The second equality is allowed because the wire is uniform, so that a length s of it will have a mass which is simply proportional to s; for example, if we halve the length of the wire we halve its mass (we have cancelled the linear density from the top and bottom lines). The quantities $x \delta s$ and δs are continuous and so the summation processes which are used for discrete quantities have to be replaced by integrals. The expression for X takes the form

$$X = \frac{\int_{-\alpha}^{\alpha} (r \cos \theta) r\, d\theta}{\int_{-\alpha}^{\alpha} r\, d\theta} = \frac{r[\sin \alpha - (-\sin \alpha)]}{[\alpha - (-\alpha)]}$$

$$= \frac{r \sin \alpha}{\alpha}.$$

We may make an elementary check on this result by assuming that the wire forms a complete circle, in which case α would take the value 180°. Then $X = 0$ and the centre of mass is positioned at $(0,0)$, at the centre of the circle, as would be expected.

If, on the other hand, the wire formed a semicircular arc, α would have the value of 90° and $X = r/(1/2\pi) = 2r/\pi$, between half and two-thirds of the radius away from the centre of the circle. (The integration of angle is expressed in terms of radians, not degrees. 2π radians is equivalent to 360°, so that 180° $= \pi$ radians and 90° $= \pi/2$ radians. π is the ratio of the circumference of a circle and its diameter.)

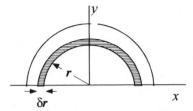

Fig. 4.6 An illustration of the axes and element shape used to find the centre of mass of a semicircular lamina.

Example 18. Locate the centre of mass of a semicircular flat plate of radius *a*.

The symmetry of the system makes the choice of axes obvious, with the origin at the centre of the circle of which the semicircle is half, and the *x*-axis along the straight edge of the semicircle. From the symmetry of the semicircle it is immediately apparent that $X = 0$. To find Y we divide the plate into semicircular elements of radius *r* and width δr, as shown in Fig. 4.6. The mass of one of these elements is proportional to its length, πr, and to its width δr, so that δm is proportional to $\pi r \delta r$. The centre of mass of the element is situated at a distance of $2r/\pi$ along the *y*-axis from the origin, and so

$$Y = \frac{\int_0^a \pi r (2r/\pi) dr}{\int_0^a \pi r \, dr} = \frac{2}{\pi} \times \frac{(a^3/3)}{(a^2/2)} = \frac{4a}{3\pi}.$$

In this last example we have introduced the statement that the mass of an element is directly proportional to its area. We may express this alternatively by saying that the lamina has a particular mass associated with its unit area. This mass per unit area is called the area density

area density of the lamina = (mass of lamina)/(area of lamina).

The lamina is of uniform thickness, say *d*, so that the area density is the mass per unit volume of the material from which the lamina is made divided by *d*. In Example 17, the mass quantity was mass per unit length of the wire, the linear density of the wire. Neither the linear density nor the area density is a generally useful quantity, because they are specific to the radius of the wire or to the thickness of the lamina respectively. The quantity 'mass per unit volume' is material specific and is the density of the material. Inevitably, the density depends on the physical state of the material, since the density of a solid is different from that of the liquid it forms on melting (remember that ice floats on water) and the densities of both are different from that of the same material in its vapour phase. A gas is compressible and has a density which depends on the conditions under which it is measured.

4.5 The human frame

The distribution of mass in the human body (on average, of course) has been established quantitatively for a variety of purposes; for example, to make the dummies suitable to simulate the occupants of vehicles in trial vehicle impacts. The results of such studies are summarized in Fig. 4.7, in which are shown and tabulated the positions of the centres of mass of the components of the human body between the hinge points which represent the effective situation of the joints in the human frame. What is remarkable about Fig. 4.7 is that the distribution of mass as shown is almost all two-dimensional. The joints all lie in the same *yz*-plane and the centres of mass of the components of the total mass, with the exception only of the feet, are all located in that same plane. This relatively simple

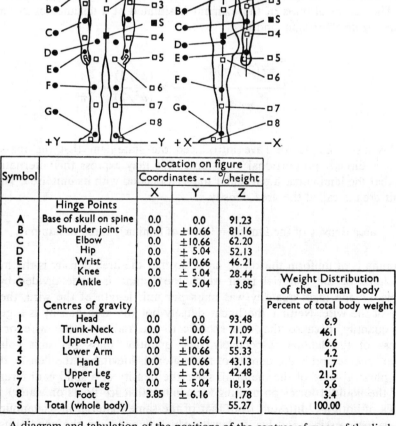

Symbol		Location on figure		
		Coordinates -- %height		
		X	Y	Z
	Hinge Points			
A	Base of skull on spine	0.0	0.0	91.23
B	Shoulder joint	0.0	±10.66	81.16
C	Elbow	0.0	±10.66	62.20
D	Hip	0.0	± 5.04	52.13
E	Wrist	0.0	±10.66	46.21
F	Knee	0.0	± 5.04	28.44
G	Ankle	0.0	± 5.04	3.85
	Centres of gravity			
1	Head	0.0	0.0	93.48
2	Trunk-Neck	0.0	0.0	71.09
3	Upper-Arm	0.0	±10.66	71.74
4	Lower Arm	0.0	±10.66	55.33
5	Hand	0.0	±10.66	43.13
6	Upper Leg	0.0	± 5.04	42.48
7	Lower Leg	0.0	± 5.04	18.19
8	Foot	3.85	± 6.16	1.78
S	Total (whole body)			55.27

Weight Distribution of the human body
Percent of total body weight
6.9
46.1
6.6
4.2
1.7
21.5
9.6
3.4
100.00

Fig. 4.7 A diagram and tabulation of the positions of the centres of mass of the limbs of an 'average' human body. After *Bioastronautics Data Book*, NASA, Washington.

arrangement in space leads to the conclusion that the body can be considered to be an arrangement of masses joined by light rods, articulated at hinge points by joints which allow movement which is restricted in a way which depends on the structure of the joint. For example, an elbow may be bent only to bring the forearm close to the bicep at the front of the upper arm, while the knee may be bent only in the direction which moves the calf towards the back of the thigh. The position of the centre of mass may be altered by repositioning the limbs and head with respect to the trunk, or by moving the trunk itself, or by both movements. By these adjustments the centre of mass of the body may be moved in the *yz*-plane in Fig. 4.7 if the movement is restricted to the *yz*-plane or to a point outside the *yz*-plane if the repositioned limb lies away from the *yz*-plane.

Example 19. A swimmer preparing to enter the water raises his arms to a horizontal position in the *yz*-plane and then lifts them so that, with straight arms, his hands touch above his head. By how much does his centre of mass move in these two actions?

The first observation we can make is that both movements are restricted to lie in the *yz*-plane, so that the centre of mass remains on the *z*-axis. Because of the symmetry of the movements the *x*- and *y*-coordinates of the swimmer's centre of mass are unchanged: $X = Y = 0$. For the *z*-coordinate we have, in effect, taken away the arms hanging vertically in Fig. 4.7 and added new arms projecting horizontally from the shoulder. The centre of mass of the arms is obtained by treating the arm components and hands as a single unit which has mass 0.125 relative to the total body mass positioned at a centre of mass situated at 0.623 of the height of the swimmer, h. The height of the shoulder joint is $0.812h$ and so the new centre of mass is positioned at

$$Z' = \frac{M[0.553 - (0.125 \times 0.623) + (0.125 \times 0.812)]h}{M},$$

where M is the total body mass, and

$$Z = Z' - Z = (0.553 + 0.125(0.812 - 0.623) - 0.553)h$$
$$= 0.024h.$$

On inspecting this procedure it is seen that, provided that we were interested only in the change in position of the centre of mass, we could have written

$$Z = 0.125\Delta l,$$

where Δl is the vertical distance through which the centre of mass of the arms has moved. When the swimmer's arms are above his head, the position of the centre of mass of the arms alters again by $0.189h$ and $Z = 0.024h$. For a person who is 1.80 m tall, this represents about 4.3 cm, so that the centre of mass has shifted (in the whole movement) by 8.6 cm vertically upwards. Roughly the same shift in the centre of mass would have taken place if the swimmer had kept his arms to his sides and stood on tiptoe. The difference between these two cases is that in the first the centre of mass moves within the body, while in the second the centre of mass is still in the same position relative to the body which has moved upwards as a whole.

Example 20. Between events, an athlete sits on the ground with trunk, head, and arms vertical and with legs stretched horizontally in front of himself. What is the position of his centre of mass? (Use the table in Fig. 4.6 for the positions of the limbs.)

The right angled bend in this problem occurs at the hip joint, so that we can derive the positions of the component centres of mass relative to that joint tabulated in x and z as follows:

Limbs	$x(h)$	$z(h)$	m/M
Feet	$0.521 - 0.018$	0.039	0.034
Upper leg	$0.521 - 0.425$	0	0.215
Lower leg	$0.521 - 0.182$	0	0.096
Trunk	0	$0.711 - 0.521$	0.461
Head	0	$0.935 - 0.521$	0.069
Arms	0	$0.623 - 0.521$	0.125

Then, for the body

$$z_B = \frac{m_A z_A + m_H z_H + m_T z_T}{m_A + m_H + m_T} = 0.197h,$$

for the leg

$$z_L = \frac{m_F z_F}{m_F + m_U + m_L} = 0.004h,$$

and

$$x_L = \frac{m_F x_F + m_U x_U + m_L x_L}{m_F + m_U + m_L} = 0.204h.$$

If we set $m_1 = m_A + m_H + m_T$ and $m_2 = m_L + m_F + m_U$, and write

$$Z_1 = z_L, \qquad Z_2 = z_B,$$

then

$$Z = \frac{m_1 Z_1 + m_2 Z_2}{m_1 + m_2} = m_1 Z_1 + m_2 Z_2,$$

since the sum of the masses in the denominator is our unit of mass, M. This leads to

$$Z = 0.130h$$

and

$$X = m_1 x_1 + m_2 x_2 = 0.131h.$$

If we assume that $h = 1.80\,\text{m}$, then $X = 23.6\,\text{cm}$ and $Z = 23.5\,\text{cm}$, so that the centre of mass of the sitting athlete is situated more or less at the corner of a square, the opposite corner of which is the base of the athlete's spine.

Exercises

4.1. A water molecule may be visualized as an oxygen atom of mass 16 to which are attached two hydrogen atoms each of mass 1. The bonds joining the hydrogen atoms to the oxygen atom are of length 0.1 nm and the angle between their directions is 105°. If the atoms can be treated as point masses and the bonds holding them together have no mass, what is the position of the centre of mass of a water molecule?

4.2. A haystack has the form of a right circular cone standing on a cylinder. The diameter of the base of the haystack is 10 m, the height of the cylinder is 5 m, and the length of the slant side of the cone is 6 m. What is the location of the centre of mass of the haystack?

4.3. A farm silo is constructed from metal sheet and has the form of a hollow cylinder of diameter 5 m and height 15 m, surmounted by a hollow spherical cap. What is the position of the centre of mass of the silo?

4.4. Masses of 5, 4, 3, 2, and 1 kg are placed on five of the vertices of a cube formed from rods of negligible mass. The masses 4, 3, and 2 are placed at the other ends of the edges which pass through the mass 5. If the mass 1 is situated at the other end of the body diagonal of the cube passing through the mass 5, what masses should be placed on the other three 'empty' vertices so that the centre of mass of the system of masses is at the centre of the cube?

4.5. A gymnast performs a handstand by placing his hands directly in front of his feet and then lifting his legs so that they are vertically above his hands. How far has his centre of mass moved in a vertical direction?

Solutions

4.1. The water molecule can be sketched schematically as shown in Fig. 4.8. Using the axes drawn in Fig. 4.8, the symmetry of the molecule gives us

$$Y = 0.$$

X is obtained from

$$X = \frac{2 \times 0.1 \times \cos 52.5°}{20}$$

$$= 0.006 \, \text{nm}.$$

The centre of mass of a water molecule is situated 0.006 nm from the centre of the oxygen atom towards the hydrogen atoms on a line which bisects the smaller angle between the bonds joining the hydrogen atoms to the oxygen atom.

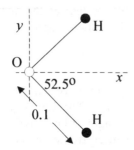

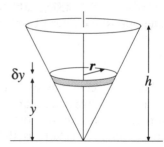

Fig. 4.8 A schematic diagram representing a water molecule.

Fig. 4.9 The choice of axes and element to find the position of the centre of mass of a cone.

4.2. Here we have to find the position of the centre of mass of a compound body, the haystack, composed from a cylinder and a cone. The rotational symmetry (about a vertical axis through the centre of the base and the apex of the stack) of the haystack means that its centre of mass is situated on the vertical axis. The centre of mass of the cylindrical base is situated at half its height, so that, for the cylindrical part of the stack $Y = 2.5\,\text{m}$, and the volume of the cylinder (which has to be multiplied by its density to obtain the mass) is $\pi a^2 h$, where a is the radius of the base and h is the height of the cylinder. The volume of the cylinder is $196.35\,\text{m}^3$. The easy way to find the centre of mass of the cone is illustrated in Fig. 4.9. If the half-angle of the cone is α, then the circular disc element drawn has a volume $\pi r^2 \delta y = (\pi y^2 \tan^2 \alpha)\,\delta y$. The element is distant y from the origin (the apex of the cone) and so

$$Y_C = \frac{\int_0^h y^3\,dy}{\int_0^h y^2\,dy}$$

$$= \frac{h^4}{4} \times \frac{3}{h^3} = \frac{3h}{4}.$$

The centre of mass of the cone is a distance of three-quarters of its height from the apex or, more relevant to this case, one-quarter of its height from its base. Since its radius is $5\,\text{m}$ and the length of its slant side is $6\,\text{m}$, its height is $3.32\,\text{m}$ and its centre of mass is $0.83\,\text{m}$ from its base, or $5.83\,\text{m}$ above the base of the haystack. The volume of the cone is

$$\int_0^h \pi y^2 \tan^2 \alpha\,dy = \pi(h^3/3)\tan^2 \alpha = (1/3)\pi a^2 h = 86.83\,\text{m}^3.$$

Relative to the ground, we have, for the compound body,

$$Y = \frac{(196.35 \times 2.5) + (86.83 \times 5.83)}{(196.35 + 86.83)} = 3.52\,\text{m}.$$

We were able to use volume instead of mass—that is, leave out the density—because the haystack is homogeneous. Its density is the same throughout, and would have cancelled out if we had taken the trouble to put it in. Such short cuts should be used with caution.

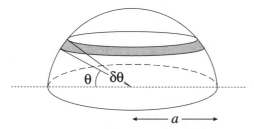

Fig. 4.10 The construction used to locate the centre of mass of a hemispherical shell.

4.3. The cylinder forming the lower part of the silo has its centre of mass at a height of 7.5 m from its base, and a volume $(\pi d)ht = 75\pi t$, where d is the diameter of its base, h is its height, and t is the thickness of its walls. Since the plate used in the construction of the silo is of uniform thickness, t may be omitted just as we have omitted the density. To find the centre of mass of the hemispherical cap we use the construction shown in Fig. 4.10. If the radius of the hemisphere is a, the 'ring' element has a volume of $a\delta\theta(2\pi a \sin \theta)$ and its centre of mass is situated at $a \cos \theta$. Then, integrating between the limits $\theta = \frac{1}{2}\pi$ and $\theta = 0$,

$$Y_h = \frac{\int 2\pi a^3 \cos \theta \sin \theta \, d\theta}{\int 2\pi a^2 \sin \theta \, d\theta}$$

$$= [(a/2)\sin^2 \theta]/1 = a/2.$$

The volume of the hemispherical shell is $2\pi a^2 t$, so that the centre of mass of the silo is situated at a height of

$$Y = \frac{(75 \times 7.5) + (17.5 \times 2 \times 2.5^2)}{(75 + 12.5)} = 8.93 \, \text{m}.$$

Here, as well as the density and thickness, we have omitted the factor π which cancels from the top and bottom lines.

4.4. Taking the cube side as unit length, and the corner at which the mass 5 is situated as the origin, the vertices of the cube and the masses placed at them are $(0,0,0)$ & 5, $(1,0,0)$ & 3, $(0,0,1)$ & 2, $(0,1,0)$ & 4, and $(1,1,1)$ & 1, and the three unoccupied vertices are $(1,1,0)$, $(0,1,1)$, and $(1,0,1)$. The centre of mass is required to be at the position $(\frac{1}{2},\frac{1}{2},\frac{1}{2})$. Using the coordinates and masses given (as shown in Fig. 4.11) we may set up a system of equations such as

$$\frac{1}{2} = X = \frac{(3 \times 1) + (1 \times 1) + (p \times 1) + (q \times 1)}{(15 + p + q + r)}$$

where p, q, and r are the unknown masses. The three equations that result

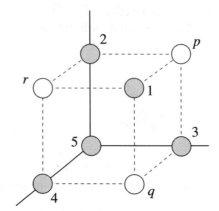

Fig. 4.11 Masses situated at the vertices of a cube.

can be solved quite easily. More briefly, we might have said that since the centre of mass lies at the centre of the cube, the mass on the upper plane face of the cube must be the same as that on the lower plane face; that is,

$$2 + 1 + r + p = 5 + 4 + 3 + q \qquad \text{or} \qquad r + p - q = 9.$$

Similarly, for the front and rear planes,

$$r + q - p = 5,$$

and for the side planes,

$$p + q - r = 7,$$

which are very quickly solved to give

$$r = 7\,\text{kg}, \qquad p = 8\,\text{kg}, \qquad q = 6\,\text{kg}.$$

The mass of 7 is situated at $(0, 1, 1)$, the mass of 8 is at $(1, 0, 1)$, and the mass of 6 is at $(1, 1, 0)$.

4.5. (The information given in Fig. 4.7 has been used in this calculation.) In the standing position the gymnast's centre of mass is at height of $0.553h$ above ground level, where h is the gymnast's height. If the gymnast balanced on the top of his head and kept his arms to his sides, he would have changed the position of his centre of mass by $h[0.553 - (1 - 0.553)] = 0.106h$, downwards. The results of his actions are slightly more complicated. In effect, he has raised his arms above his head, raising the centre of mass of his arms by δl and moving his centre of mass upwards from its initial position by $(m\,\delta l/M)$ or $0.024h$. In the inverted position the distance from the soles of his feet to the ground is $1.161h$, and his centre of mass is $0.529h$ from the soles of his feet. His centre of mass is then $0.632h$ above the ground, and has moved upwards by $0.079h$. For a gymnast 1.8 m tall, that would be about 14 cm. In the intermediate position, 'touching his toes', the gymnast's centre of mass is at its minimum height above the ground, at $0.344h$. During the whole exercise he lowers his centre of mass by $0.209h$ and then raises it by $0.228h$ to $0.632h$. The total 'down and up' movement of his centre of mass is almost $0.5h$, or 90 cm.

5
Straight line motion

5.1 Centre of mass velocity

The behaviour of particles and bodies when their positions change can be examined by considering the effect of movement of the centre of mass of the particles or of the bodies. We begin by taking a one-dimensional example in which two particles, the masses of which are m_1 and m_2, are positioned at x_1 and x_2 along an x-axis. The position of the centre of mass of the system of two particles is defined by

$$X = \frac{m_1 x_1 + m_2 x_2}{m_1 + m_2}.$$

Provided that neither m_1 nor m_2 is zero, the centre of mass lies somewhere between x_1 and x_2. The effect of motion of the two masses on the centre of mass, when they have constant velocities along the line separating them, is obtained by making the transition from position to velocity, as we did in establishing the form of the velocity vector; that is, by differentiating the expression for the position of the centre of mass with respect to time and setting the derivatives with respect to time equal to velocities, so that $dx_1/dt = v_1$, etc. This procedure defines a velocity V of the centre of mass as

$$V = \frac{m_1 v_1 + m_2 v_2}{m_1 + m_2}.$$

The velocities, V, v_1, and v_2, are all directed along the x-axis in this one-dimensional case. In a more general problem involving two or more dimensions, the equation for V given above would be an equation describing the relationship of the x-component of the velocity of the centre of mass to the x-components of the velocities of the particles. We may immediately write down expressions for the relations of the y- and z-components, add the three components to form a vector (i times the x-component, etc.) and rearrange the equation in a vector form as

$$(m_1 + m_2)V = m_1 v_1 + m_2 v_2.$$

As an alternative, we could write

$$m_1(v_1 - V) = -m_2(v_2 - V),$$

which is an expression of the equation in terms of the velocities of the particles in the frame of reference the origin of which is the centre of mass. As illustrated in Fig. 5.1, $v_1 - V = w_1$ is the velocity of m_1 with respect to the centre of mass and $v_2 - V = w_2$ that of m_2 with respect to the centre of mass, so that we may write

$$m_1 w_1 = -m_2 w_2 \qquad \text{or} \qquad m_1 w_1 + m_2 w_2 = 0;$$

that is,

$$\frac{w_{1x}}{w_{2x}} = -\frac{m_2}{m_1} = \frac{w_{1y}}{w_{2y}} = \frac{w_{1z}}{w_{2z}}.$$

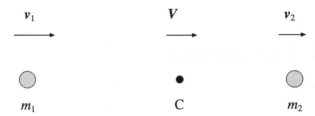

Fig. 5.1   The velocity V of the centre of mass, C, of a two particle system in which the particles have masses m_1 and m_2 and their velocities are v_1 and v_2 respectively. The velocity of m_1 with respect to C is $v_1 - V$ and that of m_2 with respect to C is $v_2 - V$.

The three ratios of the x-, y-, and z-components of the velocities, respectively, for the two particles measured with respect to their centre of mass, have the same magnitude as the inverse ratio of their masses. In this two-body problem the ratio of the two speeds is in inverse ratio to the masses of the particles.

5.2   Linear momentum

The product mw is obviously a quantity of some interest and is called the (linear) momentum, p_g, measured with respect to the centre of mass (that is, measured in the frame of reference belonging to the centre of mass). This implies that we can ascribe momenta, p_i $(= m_i v_i)$ to each component of the system. The way in which we have defined the velocity of the centre of mass tells us that V is constant provided that the v_i's are constant, so that we may write

$$p_1 + p_2 = m_1 v_1 + m_2 v_2 = \text{constant},$$

which is rather a tautology here, but will be seen to have extensive implications later.

The result obtained in the centre of mass frame of reference,

$$p_{g1} + p_{g2} = 0$$

can be extended easily to the case in which the system under consideration contains a large number of particles, giving

$$\Sigma p_{gi} = 0.$$

This means that we may talk about the linear motion of a system of particles, which might even form a solid body, in terms of the motion of the centre of mass of the system. That may seem a statement of the obvious, for if we throw an empty rigid box of internal volume, V, across the room, its sides do not move in or out relative to the centre of mass, nor do they flex, so that the walls have no linear motion with respect to the centre of mass. The empty box is not really empty, it contains air in the form of a gas. Can we deal with the gas molecules inside the box? Their position is not localized with respect to the centre of mass of the box, but their centre of mass will be at some fixed position with respect to the walls of the box. A detailed description in terms of the individual particles (the molecules) in the box would involve something of the order of 10^{25} velocity parameters per cubic metre of gas volume. Because we can describe the motion of the gas enclosed in the box in terms of the velocity of its centre of mass, or the centre of mass of the box which contains the gas, the problem has been reduced from one involving 10^{25} parameters per cubic metre to a problem involving one parameter only. The result that the linear momentum measured in the centre of mass frame of reference (the centre of mass frame, for short) is constant when the centre of mass has a uniform velocity is of great importance in reducing problems intrinsically involving a very large number of velocity parameters to an easily amenable form involving only one velocity parameter (for example, that of the centre of mass of the above-mentioned box–gas system).

5.3 Force, impulse, and reactive force

Let's now look at the two-particle system under a slightly different set of conditions. Initially, let the particles be side by side, at rest, separated only by a small explosive charge, or by a compressed spring (in this second state there must be some sort of mechanism for holding the spring in compression). In this state of rest, $v_1 = v_2 = 0$, so that the momentum of the system is zero. When the explosive charge is fired the momenta of the two particles rise almost instantaneously to constant values, because the explosion is very short-lived. The explosion has changed the state of the particles from one of rest in which there is no motion ($v_1 = v_2 = 0$) to one in which the particles move with constant velocity in a straight line. The system has been formed into a state in which the momenta of the particles, measured in the centre of mass frame, add up to zero, just as when they were initially at rest. And yet the system has altered quite radically. If we plot the momentum of one of the particles as a function of time, it will be represented almost by a step function, as in Fig. 5.2(a), because of the short lifetime of the explosion. Is the explosion likely to have a much different effect from the case in which the masses are separated by the compressed spring? The initial state and the final state can be expressed in the same terms, zero momentum and some fixed momentum, for each of the particles. The difference between the two cases lies in

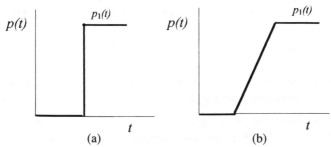

Fig. 5.2 The momentum of one of the two particles separated from each other by (a) an explosion, and (b) by the extension of a compressed spring, plotted as a function of time.

the intermediate state in which the momentum alters. This will now occupy the time interval for which the spring remains in contact with the two masses, the time taken for the spring to extend to its natural length. In the case of the masses separated by the spring, the intermediate state will last over a time interval that is relatively long, as shown in Fig. 5.2(b). If the size of the explosive charge and the elastic properties of the spring are chosen appropriately to produce the same final state for the two-particle system, we have two possible intermediate states which lead to the same final state, and it would be convenient to be able to differentiate between the two cases. The two situations depicted in Fig. 5.2 are distinguishable only by the slope of the line representing the intermediate state, so that we could use that slope to characterize the two different events. This slope, the derivative of the momentum $p_1(t)$ with respect to time, is called a force, F_1, and

$$F_1 = \frac{dp_1(t)}{dt};$$ (1a)

but, since $p_1 = m_1 v_1$, and m is constant (except under the conditions for which the theory of relativity has appreciably large effects—that is, when v is of the order of $10^7 \, \text{m s}^{-1}$), this may be written as

$$F_1 = m_1 \frac{dv_1}{dt} = m_1 v_1 \frac{dv_1}{dx_1} = m_1 \frac{d^2 x_1}{dt^2},$$ (1b)

since dv_1/dt is the acceleration of the particle 1. Equation (1a) may be integrated to give

$$\int F_1 \, dt = \int dp_1,$$

with limits on the momentum integral of $p_1(t)$ and $p_1(0)$ representing the momenta of the particle at time t and at time 0 (the initial state), so that

$$\int F_1 \, dt = p_1(t) - p_1(0).$$ (2a)

It's convenient to have a name for the integral on the left-hand side of eqn (2a). It is known as the impulse, J, of the force, F, which is applied over a time interval t. If the change in momentum is represented as δp, then eqn (2a) may be written as

$$\int F_1 \, dt = J_1 = \delta p_1 = p_1(t) - p_1(0) = m_1 v_1(t) - m_1 v_1(0).$$ (2b)

The graphical interpretation of eqn (2b) is shown in Fig. 5.3; the time variation of the force F_1 is made explicit in Fig. 5.3 by labelling the force axis $F(t)$.

We have examined the motion of one of the particles in this two-particle problem and have defined force and impulse in terms of rate of change of momentum and of change of momentum respectively. Now let's continue by looking at the system of the two particles together. In the centre of mass frame the

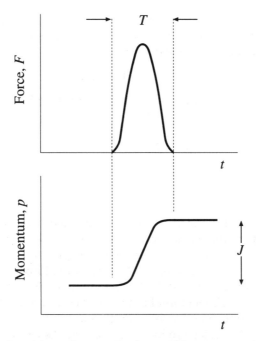

Fig. 5.3 The effect of applying a time-varying force $F(t)$ to a particle on the momentum, p, of the particle. The impulse, J, of the force is the area under the $F(t)$–t curve, or the vertical separation of the initial and final horizontal momentum lines.

momenta of the two particles have to be equal in magnitude and opposite in direction. The impulse which has acted on the particle labelled 2 is thus

$$J_2 = p_2 = -p_1 = -J_1$$

and, since the time during which the impulse acts will be the same for either particle (for example, the spring will remain in contact with either particle until it falls away from them), it is obvious that

$$F_1 = -F_2;$$

in other words, the forces acting on the two particles are equal and opposite. If the mass m_2 is imagined to become larger and larger until it is very much greater than m_1, there will still be a force F_2 acting upon it but, because of the very large value of the mass of particle 2, the velocity produced by the impulse of that force will be very small indeed. Even when that small velocity is to all intents and purposes zero, as it might be when a projectile (playing the part of particle 1) is fired vertically upwards from the surface of the earth, the force F_2 still exists and is called a reactive force, even though it has no observable effect.

5.4 Kinetic energy

If we consider one body only, acted on by a single force, its motion will be along the line of action of the force which we may define as the x-axis. This means that both the force and the velocity of the body will be directed along the x-axis so that we can rewrite Equation 1a in scalar form as

$$F = mv(dv/dx).$$

The integral form of this equation, if $v = 0$ when $x = 0$ and $v = v(x)$ at an arbitrary value of x is

$$\int_0^x F \, dx = m \int_0^{v(x)} v \, dv$$

or

$$Fx = \tfrac{1}{2} m[v(x)]^2.$$

The right-hand side of this equation is called the kinetic energy, T, of the particle. If the lower limits of the integrals were non-zero, the right-hand side of the equation would be the change in the kinetic energy, $\delta T = \tfrac{1}{2} m \, \delta(v^2)$. The left-hand side of the equation is called the work, W, done by the force. In this case the work done by the force is the product of the force and the distance through which it acts. It takes this particularly simple form because, in the case of the two particles, F_1 is directed along the x-direction; but it's easy to see that for the more general force, F, the work done is the scalar product of F and the displacement, δs, of its point of application while it is acting on the particle (or body). If we say that

$$F = i F_x + j F_y + k F_z$$

and

$$\delta s = i \, \delta x + j \, \delta y + k \, \delta z,$$

then

$$F \cdot \delta s = F_x \, \delta x + F_y \, \delta y + F_z \, \delta z$$

$$= \tfrac{1}{2} \left[m v_x^2 + m v_y^2 + m v_z^2 \right] = T.$$

Notice that the kinetic energy, as well as the work done by the force, is a scalar quantity. This property makes them distinct from impulse, momentum, force, acceleration, velocity, and position, all of which are vector quantities.

The vector quantity, p, the momentum, may be written in the form

$$p = i p_x + j p_y + k p_z,$$

so that

$$p \cdot p = |p|^2 = p^2 = p_x^2 + p_y^2 + p_z^2$$

$$= m^2 \left(v_x^2 + v_y^2 + v_z^2 \right) = 2mT$$

or

$$T = p^2/2m.$$

which is a simple relationship between momentum and kinetic energy, illustrated in Fig. 5.4.

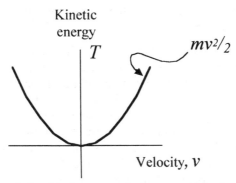

Fig. 5.4 The parabolic relationship between velocity, v, and kinetic energy, T. T is positive for negative values of v as well as for positive values of v, showing the scalar nature of T.

Example 21. Two bodies of equal mass are initially at rest in adjoining positions, separated only by an explosive charge. A mass of 1 kg is added to one of the masses and the charge is fired. The subsequent velocities of the two bodies are 2 and $3\,\mathrm{m\,s^{-1}}$. What is the mass of one of the bodies? If instead of the 1 kg mass an unknown mass, y, had been added to one of the bodies and the velocities after separation were 1.8 and $4\,\mathrm{m\,s^{-1}}$, what would be the value of the unknown mass?

After the explosive separation the momentum in the centre of mass frame of reference is unchanged. This may be visualized as a consequence of the relationship $F_1 = -F_2$—the forces acting on the two bodies are equal and opposite. Since these forces are applied for the same time, the impulse $J_1 = -J_2$. The consequence is that the total momentum of the system is still zero. Taking the mass of each body as m, this leads to $2(m + 1) = (3 \times m)$ for the first experiment; that is, $m = 2$. In the second experiment, using the same method, $y = 2.44\,\mathrm{kg}$.

Example 22. A rocket, the bodyshell of which has a mass of 100 kg, is travelling at $300\,\mathrm{m\,s^{-1}}$ relative to the ground. It contains as fuel 10 kg of hydrogen and 80 kg of oxygen which combine chemically to form water. When all this fuel has been consumed, the rocket is travelling at $2370\,\mathrm{m\,s^{-1}}$. At what speed have the water molecules been ejected from the rocket?

Let the velocity of ejection of the water molecules be V. The ejection of the water molecules is, in effect, an explosive separation of the liquid gas and the rocket shell. The total momentum of the system after this separation will be the same as the total momentum of the system before the separation took place; that is,

$$300 \times 190 = (2730 \times 100) + 90V,$$

so that $V = -2000$ and the speed of ejection is $2000\,\mathrm{m\,s^{-1}}$.

Example 23. A man of mass 75 kg stands at one end of a plank of length 5 m. The plank is at rest on a set of rollers which offer no resistance to its movement. The rollers rest on a smooth underlying surface. If the plank has a mass of 15 kg, how far will the man have moved relative to the underlying surface if he walks from one end of the plank to the other?

If there is no resistance to the movement of the plank, then the system of man and plank is an isolated system. In the absence of any force applied from outside the system, the centre of mass of the system will not move. When the man stands in his start position the centre of mass of the system (assuming the plank to be uniform so that its centre of mass lies at its centre) is 0.417 m away from the end of the plank on which the man stands towards the centre of the plank. When he is in his finishing position the centre of mass is situated at 0.417 m from the other end of the plank. The man has moved through a distance of 0.83 m relative to the surface supporting the plank. The plank has moved through a distance of $[5 - (2 \times 0.417)] = 4.167$ m.

A surface which offers no resistance to motion which takes place on it is said to be a smooth surface. The resistive effects are usually due to friction and so a smooth surface makes frictionless contact with bodies resting or moving on it. Such contacts seem desirable in practice—for example, ski jumpers would attain maximum velocity on take-off if their skis made frictionless contact with the snow—but then the slightest of side winds would give their motion an undesirable lateral component. An important point emphasized by this example is that if we have an isolated system then its centre of mass remains in the same place—that is, its velocity is zero—or it continues to move with the same velocity (which is not zero), unless there is some external agency acting to alter that velocity.

Example 24. Two cars, each of mass 1 tonne (1000 kg) are manoeuvring to park on the otherwise empty top floor of a multi-storey car park. The first has a component of velocity 3 kph parallel to the wall against which it will park and a component of velocity of 4 kph perpendicular to the same wall. The corresponding velocity components for the second car are 3 kph and 1 kph respectively. What is the angle between the momentum vectors for the two vehicles? What are the values of their momenta? Are they likely to collide?

The vectors describing the velocities are
$$v_1 = 3i + 4j, \qquad v_2 = 3i + j.$$
Since each vehicle weighs 1 tonne the momentum vectors (expressed in units of tonnes kph for convenience) are simply
$$p_1 = 3i + 4j, \qquad p_2 = 3i + j,$$
which are directed along angles defined in the coordinate system as $\tan^{-1}(4/3) = 53.13°$ and $\tan^{-1}(1/3) = 18.43°$, so that the angle between their directions is 34.7°. The magnitudes of the momentum vectors are $p_1 = 5$ tonne kph and $p_2 = 3.16$ tonne kph or, in more conventional units, 1389 kg m s^{-1} and 878 kg m s^{-1}.

A collision will take place if the relative position vector (the relative velocity vector multiplied by time) joins the initial positions of the cars and if they maintained their velocities until they collided. Since we know neither the initial positions of the cars nor whether they will maintain their velocities it is impossible to say whether they will collide or not.

Alternatively, we cannot complete the integration of the velocities of the cars to give their positions unless we know their initial positions (which are the boundary conditions of the collision problem).

Example 25. A 1 tonne experimental rocket standing on its launch pad emits its burnt fuel effectively in the form of a gaseous jet of cross-section $1 \, \text{m}^2$. The density of the gas is $10^{-1} \, \text{kg m}^{-3}$ and its velocity is $300 \, \text{ms}^{-1}$. If the gas surrenders all its momentum as it collides with the launch pad, what will be the rate of momentum transfer to the pad?

In 1 s the 'length' of the gas jet which strikes the pad is 300 m. This corresponds with a mass of $(300 \times 10^{-1}) \, \text{kg}$ of gas. In this 1 s the 30 kg of gas has lost its velocity of $300 \, \text{ms}^{-1}$, so that the rate of charge of momentum is $9000 \, \text{kg m s}^{-2}$ or 9 kN. Here we have calculated the rate of charge of momentum at the surface of the pad; in other words, the force acting on the pad. This force could be visualized as the reactive force on the surface of the earth corresponding with an equal force tending to lift the rocket off the ground. It would require a force of about 10 kN to lift a 1 tonne rocket off the ground, with the consequence that it appears that this rocket would remain firmly anchored on the launch pad. Such an observation neglects the effect of the fuel burning and being ejected. Provided that the rocket contained more than 100 kg of fuel, its mass would be reduced so that the force produced would be sufficient to cause it to leave the launch pad.

If the example is examined in terms of symbols, saying that the rocket mass is M, the jet cross-section is A, the jet velocity is v, and the jet density is ρ, then the rate of charge of momentum at the launch pad would be

$$F = \mathrm{d}p/\mathrm{d}t = A\rho v \times v = A\rho v^2,$$

as illustrated in Fig. 5.5. This force could be increased by increasing A, by increasing ρ, by increasing v, by increasing any pair chosen from them, or by increasing them all together.

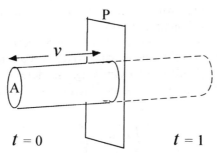

Fig. 5.5 A cylindrical jet of fluid of density ρ, cross-sectional area A, and velocity v could be visualized as composed of cylindrical segments of length v. At the time $t = 0$ an imaginary fixed plane P is placed at one end of one of these segments. One second later ($t = 1$) the segment, because it is travelling with velocity v, will have passed completely through P and be in the position shown by the 'broken-line' cylinder. The momentum transferred across P in this time interval will be $(A\rho v)v = A\rho v^2$, because the mass of the segment is $A\rho v$.

One intriguing point about this investigation of the force exerted by a jet of gas is that the argument may be inverted to provide a crude estimate of the force acting on a body travelling through a gas. Visualize a body of cross-sectional area A perpendicular to its direction of travel, moving with a velocity v through a gas. In the body's frame of reference a column of gas of area A and velocity v is impinging on it and, while we know nothing about the momentum transfer on impact of the gas molecules on the body's surface, it seems plausible to suggest that the force resisting the body's motion through the gas might well be proportional to v^2, the square of the velocity of the body through the gas.

Example 26. A sprinter leaves his blocks and attains a starting speed of $5\,\mathrm{m\,s^{-1}}$ in a time of 0.33 s. If the sprinter weighs 70 kg, what is the impulse which his legs exert when he starts from the blocks? What is the average force acting to give the sprinter this acceleration?

The change in momentum of the sprinter is from zero to $350\,\mathrm{kg\,m\,s^{-1}}$, so that the impulse to which he is subjected is $J = 350\,\mathrm{kg\,m\,s^{-1}}$. Since this impulse occupies a time of 0.33 s, the average force acting in that time is

$$F = J/t = 350/0.33 = 1050\,\mathrm{N}.$$

The acceleration of the athlete will be $a = F/m = 1050/70 = 15\,\mathrm{m\,s^{-2}}$, a result which could have come as easily from $a = v/t = 5/0.33 = 15\,\mathrm{m\,s^{-2}}$.

Example 27. A baseball pitcher with a fast arm pitches a baseball so that it is travelling at $40\,\mathrm{m\,s^{-1}}$ along a horizontal line when it is hit by the batter. The baseball, which has a mass of 170 g, leaves the bat with a speed of $60\,\mathrm{m\,s^{-1}}$ along a direction which is 30° above the horizontal passing over the pitcher's head. If the time of contact between bat and ball is 15 ms, what is the average force exerted on the bat?

The impact between bat and ball lasts for a short, finite, time interval, from when the ball first touches the bat until it leaves the bat. In the interval between these two extremes the force will rise to a maximum value and then decrease, as illustrated in Fig. 5.6. Although the force–time curve in Fig. 5.6 is drawn as approximately symmetrical, it could be quite unsymmetrical. The detail of the force–time curve depends on the way in which the bat–ball interaction takes place. In order to avoid the difficulties of examining the detail of this interaction, we average the force over the time interval in which the bat is in contact with the ball using the relationship.

$$\frac{\int F\,dt}{\int dt} = \bar{F}$$

or

$$\int F\,dt = \bar{F}t = \delta p$$

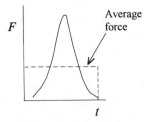

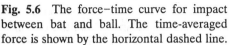

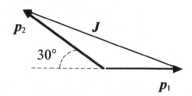

Fig. 5.6 The force–time curve for impact between bat and ball. The time-averaged force is shown by the horizontal dashed line.

Fig. 5.7 The momentum vector diagram for the bat–ball interaction.

where $\bar{F}$ is the average force, generally much less than the maximum force (they would be equal only if the 'bell' shaped curve in Fig. 5.6 had a top hat form with vertical sides and a horizontal top). This average force is a 'guesstimate' which provides a rough pointer to the magnitude of the maximum force exerted by ball on bat; it is no more than a convenient fiction, enabling a rough estimate of the force to be made from the observed change in momentum.

The momentum vector diagram used to find the momentum change is shown in Fig. 5.7. The initial momentum of the ball is

$$|\mathbf{p}_1| = |m\mathbf{v}_1| = 0.17 \times 40 = 6.8,$$

directed along a horizontal line towards the batter. Its final momentum is $p_2 = 0.17 \times 60 = 10.2$, directed along a line inclined at an angle of 30° above the horizontal towards the pitcher. Application of the cosine rule

$$(\delta p)^2 = p_1^2 + p_2^2 + 2p_1 p_2 \cos 30°$$

leads to $\delta p = 16.44\,\text{N s}$, so that the average force $|\bar{F}| = 16.44/0.015 = 1096\,\text{N}$. The maximum force will be well above this average in general, depending on the detail of the interaction of bat and ball.

The everyday meaning of the word 'impulse' misleadingly implies that impulsive forces are associated only with sudden impacts. In fact, a force of 5 N acting on a particle for 20 s would give rise to an impulse of 100 N s. If the particle were moving freely its momentum change during this period would be 100 N s, the same as the impulse, of course. The impulse of a force is the change in momentum of the (freely moving) body to which it is applied during the time for which it is applied. While its momentum is changing the body is being subjected to an impulse. When its momentum has changed, the component of momentum along the line of the impulse has altered by the magnitude of the impulse.

Example 28. A tennis ball of mass 0.1 kg is kicked so that it receives an impulse of 2 N s. What momentum change is observed (i) if the ball is at rest initially, and (ii) if it has an initial velocity of $10\,\text{m s}^{-1}$ in the direction of the impulse? What are the corresponding kinetic energy changes? What difference would it make if the initial velocity of $10\,\text{m s}^{-1}$ and the impulse were directed at right angles to one another?

The momentum change is the same in either case, $\delta p = 2$, so that whether or not the ball is initially stationary $\delta v = 20\,\mathrm{m\,s^{-1}}$. In case (i) the ball will acquire a velocity of $20\,\mathrm{m\,s^{-1}}$ and its kinetic energy will change from zero to $(0.5 \times 0.1 \times 20^2) = 20\,\mathrm{J}$. In case (ii) the ball will have its velocity increased by $20\,\mathrm{m\,s^{-1}}$ from $10\,\mathrm{m\,s^{-1}}$ to $30\,\mathrm{m\,s^{-1}}$; its kinetic energy increases from $5\,\mathrm{J}$ to $45\,\mathrm{J}$, an increase of $40\,\mathrm{J}$.

Why is it that the changes in momentum (or velocity) are the same in both cases while the kinetic energy changes are quite different? The answer is simply related to the distance through which the ball travels whilst the force is acting upon it. As illustration, let the impulse be the result of a constant force of $20\,\mathrm{N}$ applied to the ball for $0.1\,\mathrm{s}$. The acceleration from rest is $a = 20/0.1 = 200\,\mathrm{m\,s^{-2}}$ and the distance travelled by the ball while the force is applied is $x = \frac{1}{2}at^2 = 1\,\mathrm{m}$. The work done by the force, the increase in the kinetic energy of the ball, is $20\,\mathrm{J}$. In the second case the acceleration of the ball is the same but the distance travelled while the force is applied is $x = v_0 t + \frac{1}{2}at^2 = 2\,\mathrm{m}$. The work done by the force is $40\,\mathrm{J}$, equal to the change in the kinetic energy.

When the impulse and velocity are perpendicular to one another the momentum vector triangle is right angled, with its short sides of length 1 and $2\,\mathrm{N\,s}$. The resultant momentum is $2.236\,\mathrm{N\,s}$, and the final velocity is $22.36\,\mathrm{m\,s^{-1}}$. The final kinetic energy is then $25\,\mathrm{J}$ so that the change in kinetic energy, $20\,\mathrm{J}$, is the same as when the ball was initially at rest. If the direction of the initial velocity is the x-direction, this is because the impulse alters only the y-component of the velocity which was initially zero and becomes $20\,\mathrm{m\,s^{-1}}$. The kinetic energy associated with the x-component of the velocity is unaltered because the force which constitutes the impulse acts at right angles to its direction; since this force has no component in the x-direction the product of the x-component of the force and the distance the body moves in the x-direction while the force is applied must be zero,

$$F_x\,\delta x = \tfrac{1}{2}m\,\delta(v_x^2) = 0.$$

5.4.1 No work forces

Example 29. In the solar system the planet Venus travels round the sun in an orbit which, to a first approximation, is a circle of radius $1.08 \times 10^8\,\mathrm{km}$. The mass of Venus is $4.9 \times 10^{24}\,\mathrm{kg}$, and its 'year' is 224 days. Assuming that the speed of the planet in its orbit is constant, find the force acting on it to keep it in its orbit. How much work is done by this force, and at what rate does it work?

The acceleration of a particle of mass m travelling at a constant speed v along a circular arc of radius r is $a = v^2/r$. The force required to produce this acceleration is $F = ma = mv^2/r$. To evaluate F we need to know v, which is simply $v = 2\pi r/T$, the circumference of the orbit divided by the time taken to travel around the orbit. Using the data provided, $v = 3.5\,\mathrm{m\,s^{-1}}$, and $F = 5.57 \times 10^{22}\,\mathrm{N}$. In comparison with the magnitude of forces experienced in everyday existence this is almost inconceivably large. The force is directed towards the centre of the orbit, always perpendicular to the direction of the velocity—the one is normal, the other tangential, to the orbital path. Because the force and the velocity are always directed at right angles to one another, the force does no work and its rate of working is zero. The force does no work because its function is solely to alter the direction of the velocity vector of the planet, the magnitude of the velocity is unaltered by the force. Alternatively, the orbit is unaltered in time and the kinetic energy of the planet is constant—no work is done on the planet; in other words, the force does no work.

The rate at which work is done by a force is

$$P = \frac{d}{dt}(F \cdot r) = \frac{dW}{dt}$$

which, for a constant force is $F \cdot (dr/dt) = F \cdot v$. Power, measured in watts (or Js^{-1}), is a convenient quantity which tells us the rate at which work is produced by a machine or the rate at which work is consumed by a machine. As example, the engine in a compact car can produce a power of around 40 kW, although this is just as likely to be expressed in terms of horsepower, hp (1 hp = 746 W). A domestic television set requires about 100 W to make it function. The power output from a given machine is often not constant in time, in which case the work done by the machine is

$$W = \int P(t)dt$$

which leads, of course, to the possibility of defining an average power.

5.5 Pressure and vector product

Example 30. A jet of water produced by a hosepipe has a cross-sectional area of $20 \, cm^2$. The jet is directed so that it is incident normally on a vertical wall. If the water jet has a velocity of $12 \, ms^{-1}$ and loses all its velocity in impact with the wall, what force does the jet exert on the wall? The density of water is $10^3 \, kg \, m^{-3}$.

The volume of water arriving per second at the wall is the volume of a cylinder of cross-section $2 \times 10^{-3} \, m^2$ and of length 12 m; that is,

$$V = 2 \times 10^{-3} \times 12 = 2.4 \times 10^{-2} \, m^3.$$

The mass of water arriving at the wall per second is

$$m = 2.4 \times 10^{-2} \times 10^3 = 24 \, kg \, s^{-1}.$$

The rate of change of momentum, the momentum change per second of the water in the jet, is $mv = (24 \times 12) \, N$. This is the force exerted by the water jet, so that $F = 288 \, N$.

An alternative way of expressing this result would be to state it in terms of the pressure, p^*, exerted by the jet on the wall. Pressure, p, is defined as normal force per unit area, in this case for unit area of the jet, so that

$$p = 288/(2 \times 10^{-3}) = 1.44 \times 10^5 \, N \, m^{-2}, \quad \text{or} \quad 1.44 \times 10^5 \, Pa.$$

---

* In this example, the symbol p represents pressure, rather than momentum.

The pressure exerted by the atmosphere at the surface of the earth is 1.013×10^5 Pa, so that the result could be expressed as 1.42 atmospheres. The pressure exerted by rainfall on a flat roof may be calculated because the rainfall may be visualized as a water jet with a relatively small volume per unit area. A 1 cm precipitation in 1 h represents a very heavy rainfall, so that the average pressure acting on the roof is, in general, very small indeed. On the other hand, a cylindrically shaped raindrop of 2 mm diameter and 1 mm long parallel to its direction of motion falling vertically with a velocity of $10 \, \text{ms}^{-1}$ impinging on a flat roof suffers a momentum change of 10^{-5} N s in a time of 10^{-4} s; that is, it exerts a force of 0.1 N over an area of 10^{-6} m^2. The localized pressure exerted by the raindrop is 10^5 Pa, roughly one atmosphere. A similar raindrop falling with a velocity of $40 \, \text{ms}^{-1}$ would exert a localized pressure of around 4 atmospheres.

Introducing the idea of pressure in this way incorrectly implies that pressure acts in a particular direction and is a vector quantity. This apparent directionality of pressure is a consequence of introducing it by considering a unidirectional flow of fluid. The pressure, p, exerted by a fluid is the normal force per unit area acting on a surface immersed in, or containing, the fluid. If we inflate a small balloon the pressure exerted by the layer of gas at its surface must be the same as the pressure exerted by the layer of gas lying next to this outermost layer. In its turn the pressure exerted by this second layer must be the same as that exerted by a third layer, and so on. The pressure is uniform throughout the volume of the balloon; it acts in all directions, in particular out towards all parts of the surface of the balloon, and cannot be described as a vector quantity. It must be a scalar quantity. If pressure is force per unit area then we can say that force is the product of pressure and area. Because force is a vector quantity and we can equate a vector only with another vector, we must be able to represent an area A by a vector A, of magnitude A, directed normal to the plane containing the area, so that

$$F = pA.$$

In the vector representation of area the area vector must be aligned with the normal to the area in question, since F is a normal force and A and F must be collinear. To find how A can be represented by a vector we construct a parallelogram from two vectors r_1 and r_2 in directions which differ by an angle θ, as shown in Fig. 5.8. The area of such a figure is $A = r_1 r_2 \sin \theta$, and can be visualized as a vector A directed perpendicular to the plane and of magnitude A.

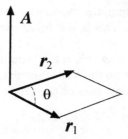

Fig. 5.8 The representation of an area A, defined by the vectors r_1 and r_2, by a vector A directed normal to the plane containing r_1 and r_2.

These observations lead us to define a new form of vector multiplication in the same sort of way that we defined the scalar product of two vectors. This is the multiplication of two vectors, r_1 and r_2, which produces a vector, here called A, which is directed perpendicular to the plane in which r_1 and r_2 lie and which has a magnitude $A = r_1 r_2 \sin \theta$. This form of multiplication of vectors is called vector multiplication and gives the vector product of the two vectors. We represent the vector product of two vectors as

$$r_1 \times r_2 = A.$$

There's an ambiguity in this definition, because A could point either up from the surface, A_u, or down into the surface, A_d. Since these two vectors will have the same magnitude it's obvious that $A_u = -A_d$. This difficulty is resolved by writing the vector products defining these two A's as

$$r_1 \times r_2 = A_u \qquad \text{and} \qquad r_2 \times r_1 = A_d,$$

in which case the order of multiplication becomes significant, because

$$r_1 \times r_2 \neq r_2 \times r_1 \qquad \text{or} \qquad r_1 \times r_2 - r_2 \times r_1 \neq 0.$$

Vector products do not commute as do scalar products $[(r_1 \cdot r_2) - (r_2 \cdot r_1) = 0]$. If we have a vector product $A = r_1 \times r_2$, then the direction of A is the direction (perpendicular to the plane containing r_1 and r_2) in which a right-hand screw would advance if its head were rotated from r_1 to r_2. If the vector product is written in the other order, the screw head would be rotated from r_2 to r_1 and the screw would advance in the opposite direction.

Since the magnitude of the vector product depends on $\sin \theta$ (where θ is the angle between the vectors) the vector product of parallel vectors is zero. This means that if we multiply two vectors $a = i a_x + j a_y + k a_z$ and $b = i b_x + j b_y + k b_z$ as a vector product, then

$$a \times b = i(a_y b_z - a_z b_y) + j(a_z b_x - a_x b_z) + k(a_x b_y - a_y b_x),$$

since $i \times j = k$, etc., and $i \times i = 0$, etc. (Strictly, $i \times i$ should be equal to the null vector whose magnitude is zero and we should have written $|i \times i| = 0$.)

5.6 Superposition of forces

From the way we have defined force it is obvious that if a system of forces F_1, F_2, F_3, etc. act simultaneously on a body then we may say that their effect, so far as the motion of the centre of mass is concerned, is the same as if a single force F acted on the body, where

$$F = \sum_i F_i.$$

Forces may be added vectorially to give a single resultant force. A body might have several forces acting on it and yet suffer no acceleration because the forces nullify each other; in other words, they add vectorially to zero. Thus a ship sailing with constant velocity needs some force to nullify the drag forces produced by the water in which its hull is immersed. When the vectors representing these forces are equal, there will be no net force acting on the ship and it will maintain a constant velocity.

This principle of superposition of forces enables us to make the observation that when the vectors representing the forces acting on a body sum to zero the momentum of the body's centre of mass remains constant. Alternatively, we may say that if the centre of mass of the body has constant momentum associated with it the sum of the forces acting on it must be zero. In the case in which the momentum of the centre of mass is zero, the body is static and the principle of superposition of forces enables us to develop the sub-group of mechanics known as statics.

Exercises

5.1. A compact car has a mass of 1 tonne. What would be the average force required to accelerate it from rest to $25 \, \text{m s}^{-1}$ in 5 s? What would be the power involved in this acceleration?

5.2. When the engines are stopped on a ship of mass 10 000 tonnes it slows from a speed of 12 kph to 10 kph in a distance of 90 m. If the resistance to the ship's motion is independent of velocity, what is its value? If 12 kph is the maximum speed at which the ship can travel, make an estimate of the power of its engines.

5.3. A hammer of mass 1 kg is used to drive a nail of mass 30 g into a block of wood. The hammer head is moving with a velocity of $6 \, \text{m s}^{-1}$ just before it strikes the nail, which then penetrates 2.5 cm into the wood. Find the velocity of the nail and hammer just after impact. What is the percentage loss of kinetic energy in the impact? For what time is the nail in motion through the wood, and what is the average force of resistance of the wood to the motion of the nail?

5.4. A cannon of mass 500 kg is used to fire a cannonball of mass 8 kg in a horizontal direction. If the muzzle velocity of the ball is $500 \, \text{m s}^{-1}$, with what velocity does the cannon recoil? What is the kinetic energy of the cannon–cannonball system? If the distance travelled along the bore of the cannon by the ball is 2.1 m, what is the average force acting on the shot? How far will the cannon have moved when the shot leaves its muzzle?

5.5. The water carried by an iron pipe of internal diameter 5 cm moves with a velocity of $4.8 \, \text{m s}^{-1}$. If there is a right angled bend in the pipe, what is the force acting on the pipe at the bend?

5.6. In making a steel stamping, a mass of 100 kg attains a velocity of $4.4 \, \text{m s}^{-1}$ by the time it makes contact with the steel. If the mass travels a further 1.2 cm before coming to rest, what is the average resistance offered to its motion by the steel? What impulse is given to the steel, what work has been done on it, and what is the average power exerted on it?

Solutions

5.1. We have the following:

$$mv = 25 \times 10^3, \quad mv/t = 5 \times 10^3 = 5\,\mathrm{kN},$$

$$W = 5000x, \quad v = at, \quad a = 5\,\mathrm{m\,s}^{-2}, \quad x = \tfrac{1}{2}at^2 = 62.5\,\mathrm{m},$$

$$W = 312.5\,\mathrm{kJ},$$

$$P = W/t = 62.5\,\mathrm{kW}.$$

Alternatively, the kinetic energy of the car, $T = \tfrac{1}{2}mv^2 = 312.5\,\mathrm{kJ}$ after its 5 s of acceleration. This is the work done by the engine of the car at an average rate of 62.5 kW.

A commonly used measure of the power produced by mechanical engines is the horsepower, 1 hp = 746 watts. By this measure the power of the family car is about 84 hp.

An objection which might be made against using a constant force approximation in examining the performance of a car is that real cars are far more constant power systems than they are constant force systems. The automatic gear changing on Formula One racing cars is arranged so that the power output of the engine remains as nearly as possible at its maximum value, so that the performance of the car is optimized. We may look at the behaviour of this family car to see how the assumption that the car produces a constant power output effects the result we obtained using the constant force approximation. We begin with the statement that power is the rate at which work is done which is also the rate of change of kinetic energy; in other words,

$$P = \frac{\mathrm{d}}{\mathrm{d}t}\left(\tfrac{1}{2}mv^2\right) = mv\,\frac{\mathrm{d}v}{\mathrm{d}t}$$

so that, on integration,

$$P\int \mathrm{d}t = m\int v\,\mathrm{d}v$$

or

$$Pt/m = \tfrac{1}{2}v^2,$$

provided that $v = 0$ when $t = 0$. Then

$$\mathrm{d}x/\mathrm{d}t = (2Pt/m)^{1/2}$$

and, on integration,

$$x = \tfrac{2}{3}(2Pt^3/m)^{1/2}$$

if $x = 0$ when $t = 0$.

The interesting point about this result is that it shows the importance of the power to mass ratio, P/m. Substituting in the numbers from the problem leads to a power value of 62.5 kW, just as in the constant force approximation. The change in approximation becomes important only when the question is taken past the very simple form in which it was posed.

5.2. We need to convert the speeds from kph to ms^{-1} to be consistent with our system of units. 1 kph is the same as 1000 m per 3600 s, i.e., 0.278 ms^{-1} so that

$$12\,\text{kph} = 3.333\,\text{m s}^{-1}, \qquad 10\,\text{kph} = 2.778\,\text{m s}^{-1}.$$

In slowing down, the ship loses kinetic energy:

$$T = (0.5 \times 10^7)(3.333^2 - 2.778^2) = 16.96\,\text{MJ},$$

so that

$$F = T/x = (16.96/90) \times 10^6 = 1.88 \times 10^5\,\text{N}.$$

The way in which we have defined force as a vector quantity implies that if we have two forces, F_1 and F_2, then the results of applying them simultaneously to a body is that the body is subjected to a force

$$F = F_1 + F_2$$

or, more generally, for several forces,

$$F = \sum_i F_i.$$

The force resulting from the simultaneous application of several forces is simply the vector sum of the forces. Vector diagrams for forces acting simultaneously (on the same body) may be constructed to establish the magnitude and direction of a single equivalent force which, often, is called the resultant of the forces from which it is derived. If the forces are collinear the sum takes on a particularly simple form, involving only their magnitudes; for example, if equal and oppositely directed forces are acting on a body the net force (the resultant force) acting on it is zero.

This observation is consistent with the vectorial definitions of position, velocity, and acceleration which we made earlier. Our ability to form a simple vector combination of forces in this way is dignified by the title 'the principle of superposition of forces'.

The ability to add up forces in an amenable way makes it possible to form an estimate of the force exerted by the ship's propulsive mechanism, because it must be equal to the resistive force when the ship is travelling with constant velocity. At constant speed the velocity does not alter so that the net force must be zero. The power output of the engines is the resistive force multiplied by the velocity; that is, at 12 kph $P = (1.88 \times 10^5 \times 3.333)\,\text{J}$ —in other words the engine power is 628 kW, or 842 hp.

5.3. Conservation of momentum requires that

$$Mv = (M + m)v_1$$

that is,

$$v_1 = 6 \times (1/1.03) = 5.825\,\text{m s}^{-1}.$$

The kinetic energy equation is

$$T = \tfrac{1}{2}Mv^2 = \tfrac{1}{2}(M + m)v_1^2 + \delta T,$$

where δT is the loss of kinetic energy. The fractional change in energy is

$$\delta T/T = 0.9425(1 + m/M) = 0.029,$$

so that the percentage loss of energy is 2.9%.

The average retardation is $a = v_1^2/2x = 33.93/0.5 = 67.86\,\mathrm{m\,s^{-1}}$. Then $v = at$ gives $t = 8.58\,\mathrm{ms}$. The momentum change for the nail is $5.825 \times 0.03 = 0.175\,\mathrm{N\,s}$, and $F = \delta p/t = 20.36\,\mathrm{N}$; or, the loss of kinetic energy of the nail is $0.5 \times 0.03 \times 33.93 = 0.509\,\mathrm{J}$ and the force is $0.509/0.025 = 20.36\,\mathrm{N}$.

5.4. Relative to the centre of mass of the system of the cannon and cannonball, we may write

$$mv_1 = Mv_2$$

in other words,

$$v_2 = 8 \times 500/500 = 8\,\mathrm{m\,s^{-1}}.$$

Since there are no external forces acting on the system, the centre of mass does not move and the kinetic energy is

$$T = 0.5[(8 \times 500^2) + (500 \times 8^2)] = 1.016\,\mathrm{MJ}.$$

The average force, $F = T/x = (1.016 \times 10^6)/2.1 = 4.84 \times 10^5\,\mathrm{N}$. The same force acts on the cannon, so that its acceleration is

$$a = F/M = 967.6\,\mathrm{m\,s^{-2}}, \qquad v^2 = 2ax, \qquad x = 64/(967.6 \times 2) = 3.3\,\mathrm{cm}.$$

5.5. The flow of water is maintained around the right-angled bend—the rate at which it arrives at the bend must be the same as the rate at which it leaves the bend. Since the bore of the pipe is uniform, the water leaving the bend travels with the same speed as the water approaching the bend. The magnitudes of the velocities of approach and leaving are the same. The direction of the velocity vector has been altered by 90° in passing round the bend, so that the change in velocity at the bend is $\delta v = 4.8\sqrt{2}\ \mathrm{m\,s^{-1}}$. The mass of water arriving at the bend per second is

$$\pi(0.025)^2 \times 4.8 \times 10^3 = 9.42\,\mathrm{kg\,s^{-1}}.$$

The rate of momentum change is $9.42 \times 4.8 \times \sqrt{2} = 64\,\mathrm{N}$, which is the force at the bend.

The momentum change of the water is greater than it would be had it simply stopped when it arrived at the bend. The force at the bend is larger than would be the force exerted by an equivalent jet of water losing its velocity by impact on a wall. Notice that the kinetic energy of the flow is not altered by the change in direction. Once again, we have a 'no work' force changing the direction of the velocity vector because the pipe, fixed in place by its supports, does not move. If it were free to move, it would do so in reaction to the force, and work would be done.

5.6. The retardation, $a = v^2/2x = 806.6\,\mathrm{m\,s^{-1}}$. The force, $F = ma = 80.7\,\mathrm{kN}$. The time of retardation is derived from $x = \frac{1}{2}at^2$, giving $t = 5.45\,\mathrm{ms}$. The impulse is then $Ft = 440\,\mathrm{N\,s}$, the work done is $W = Fx = 968.4\,\mathrm{J}$, and the power exerted on the stamping is $P = W/t = 178\,\mathrm{kW}$.

6
Energy conservation

6.1 Elastic and inelastic collisions

If we have a system composed of two masses, m_1 and m_2, moving with velocities v_1 and v_2, then the kinetic energy of the system is

$$T = \tfrac{1}{2}\left[m_1 v_1^2 + m_2 v_2^2\right].$$

The kinetic energy associated with the movement of the centre of mass of the system is

$$T_G = \tfrac{1}{2}mV^2 = \tfrac{1}{2}(m_1 + m_2)\left[\frac{m_1 v_1 + m_2 v_2}{m_1 + m_2}\right]^2$$

$$= \frac{(m_1^2 v_1^2 + m_2^2 v_2^2 + 2m_1 m_2 v_1 v_2)}{2(m_1 + m_2)},$$

which certainly is not the same as T. The difference between these two energies is

$$T_G - T = \tfrac{1}{2}\left\{\frac{(m_1 v_1 + m_2 v_2)^2}{m_1 + m_2} - m_1 v_1^2 - m_2 v_2^2\right\}$$

$$= \frac{m_1 m_2}{2(m_1 + m_2)}(v_1 - v_2)^2.$$

The kinetic energy of the system may thus be written as

$$T = T_G + \frac{m_1 m_2}{2(m_1 + m_2)}(v_1 - v_2)^2$$

an expression in which $v_1 - v_2$ is readily identified as the relative velocity of the two bodies.

If the two particles move collinearly with appropriate velocities they will collide and their velocities will be altered by the impulse of the collision. In the absence of any external agency, the velocity of the centre of mass will be unaltered. The change, δT, in kinetic energy of the system will be proportional to the difference between the squares of the relative velocities before and after the collision; that is,

$$\delta T = \frac{m_1 m_2}{2(m_1 + m_2)}(v_{r1}^2 - v_{r2}^2),$$

where v_{r1} and v_{r2} are the relative velocities of the bodies before and after collision respectively. There will be an energy loss on collision unless $v_{r1} = v_{r2}$, in which case $\delta T = 0$. When the energy loss is zero the collision is an elastic collision. When there is an energy loss the collision is inelastic. Whether the collision is elastic or inelastic, momentum is conserved (the total momentum remains constant) unless there is some external force acting on the system.

The collision of two billiard (snooker or pool) balls provides a straightforward example of an inelastic collision which may be described in terms of the ratio of the relative velocities of the balls after and before collision, the coefficient of restitution e $(=v_{r2}/v_{r1})$: $e < 1$ when kinetic energy is lost in the collision while $e = 1$ for elastic collision. This quantity could be (and sometimes is) called 'coefficient of elasticity' or perhaps 'coefficient of reflection'. These alternatives are not much used because of the possibility of confusion arising, for example, from the way in which light is reflected at surfaces (the loss of energy of light reflected at a surface is related to $(1 - r^2)$, where, here, (and only here), r^2 is the reflecting power of the surface). Relating the two relative velocities in this way leads to the expression

$$\delta T = \frac{m_1 m_2}{2(m_1 + m_2)} v_{r1}^2 (1 - e^2).$$

Evidence for the inelasticity of the collision comes from the click audible when the balls collide. Some of the kinetic energy of the balls has been transformed into sound energy. Sound energy is kinetic energy of air molecules, so that the balls have surrendered some of their kinetic energy to the surrounding atmosphere, the air. In addition, there might be an increase in the internal energy of the balls—their constituent molecules would have their kinetic energies increased (at the expense of the kinetic energies of the balls) in the collision. This would be manifested by a very small increase in their temperatures. Part of the kinetic energy of motion of the balls is transformed in the inelastic collision to sound energy and thermal energy (heat). The First Law of Thermodynamics, the law of conservation of energy, tells us that if kinetic energy—in this case of linear (or translational) motion—is lost, it must be transformed to another form of energy.

6.2 Energy conversion

The collision of atomic and nuclear particles provides further good examples of collision processes. The Franck–Hertz experiment, which provided clear evidence for the quantization of atomic energy levels, depended on the loss of kinetic energy by electrons during collisions with atoms. This 'lost' energy was transformed into internal energy of some of the atoms which were left in an excited state. These excited atoms remained in this state until they returned to their initial (ground) state by converting their surplus internal energy into light energy which they emitted. Elastic and inelastic collisions occur as well in the techniques of X-ray and neutron diffraction. For X-rays the elastic scattering gives rise to a diffraction pattern (Bragg scattering), while inelastic scattering produces an increase in the wavelength of the scattered rays (Compton scattering). The interpretation of the results of inelastic scattering experiments depends on the first law of thermodynamics which might be stated, 'the total energy of the universe is constant'. When one form of energy appears to be lost it has been transformed into an equivalent quantity of other forms of energy. Accepting the truth of the First Law of Thermodynamics means rejecting the possibility of making a perpetual motion machine which produces more energy than it consumes.

6.3 Potential energy and total energy

The idea that there are forms of energy that differ from kinetic energy opens a window to a new world in mechanics. As illustration, take the case of a rubber band which can be stretched quite easily. When it is released it snaps back into its original unstretched form, at a rate which is related to its extension from its natural length when it is in its stretched state. The work done by the force stretching the band has been stored in the band and is transformed to kinetic energy when the force is removed suddenly. This stored energy is called potential energy because it is manifested only when it is transformed to another form of energy—it provides the band with the potential to revert to its original form. Another illustration of potential energy is the case of a high board diver. The diver climbs from pool level to the board level and stands still before projecting himself outwards (and upwards) from the board. Once he has left the support of the board he dives with increasing velocity towards the water surface some distance below the board. At the height of the board he must have had a higher potential energy than he had at pool level—the process of diving has converted that potential energy to kinetic energy of falling. For argument's sake, say that when the diver stands on the board his potential energy relative to his potential energy at pool level is V_b. When he enters the water this extra potential energy will all have been converted to kinetic energy. If his velocity at pool level is v_p and his mass is m, then

$$T_p = \tfrac{1}{2}mv_p^2.$$

During his descent the diver is gaining kinetic energy at the expense of his potential energy and, at a distance z below board level,

$$T + V = \tfrac{1}{2}m[v(z)]^2 + V(z)$$

where $v(z)$ is his velocity and $V(z)$ his potential energy at the position z. His total energy. H, is the sum of his kinetic energy and his potential energy, a statement which is written often as

$$H = T + V$$

where H, T, and V represent total, kinetic, and potential energies, respectively. T is a function of velocity, and V is a function of position.

If we make the assumption that the diver is subjected to a constant force when he is falling during his dive, we may express his acceleration as $a = F/m$, where m is his mass. This acceleration is found to be constant for decreases in height which are not very large and has the value $9.81\,\mathrm{m\,s^{-2}}$. There is a force acting on the falling body directed towards the surface of the earth giving rise to this acceleration, the acceleration due to gravity, generally symbolized by g, with the value

$g = 9.81 \, \mathrm{m\,s^{-2}}$. It is not necessary at this stage to enquire into the origin of this force; it is sufficient just to note its existence and say that it is the consequence of gravitational attraction. If this force is acting on the diver it will act also on a spectator at ground level. The spectator is subjected to a force which is equivalent to that experienced by the diver, but the reactive force at the solid surface of the earth nullifies it. Since no net force acts on the spectator he does not fall through the surface, but if he wants to change his elevation he will have to do work against that force, or work will be done by the force on him. This force which, in effect, keeps the spectator's feet on the ground, is $F = m_1 g$, where m_1 is his mass. This force is the weight of the spectator and the force of reaction which nullifies the weight is often called the contact force. In general, for bodies, the product of mass and the acceleration due to gravity is the weight of the body. The weight of a body will be constant provided that its mass does not change and that g remains constant. Weight is measured in newtons; the common usage, for example, that 'a heavyweight boxer weighs 100 kg' means that the boxer has a weight of 981 N and that his mass is 100 kg.

In the example involving the diver, when he has fallen through a distance z the work which has been done on him by the force which 'attracts' him towards pool level is mgz. This work is manifested as increased kinetic energy, so that

$$\tfrac{1}{2}m[v(z)]^2 = F \times z = mgz.$$

If the board has an elevation, h, we may introduce a new variable, height above pool level y, in place of z, so that $z = h - y$ and

$$\tfrac{1}{2}m[v(y)]^2 = mg(h - y)$$

or

$$\tfrac{1}{2}m[v(y)]^2 + mgy = mgh.$$

The right-hand side of this equation is constant and has the dimensions of energy, while the terms on the left-hand side consist of variables but have the same dimensions, of energy. This leads to the identification of mgh with the initial total energy and $mgy = V(y)$ as the potential energy of the diver at height y above pool level. Notice that $V(y)$ increases with increasing elevation while the direction of F is towards decreasing y. This leads us to the proposition that in a one-dimensional system in which a force may be represented by a potential field

$$F_y = -\frac{\mathrm{d}}{\mathrm{d}y}[V(y)]$$

or, force is 'minus the gradient of the potential energy'. This relationship is illustrated for the case of the diver in Fig. 6.1.

The increase in the diver's kinetic energy during his dive is equal to the decrease in his potential energy. His total mechanical energy remains constant all the time and his potential energy depends only on y, his height above pool level. This means that the noise and any splash that he makes on entering the water—the loss of energy on entering the water—is a loss taken entirely from his kinetic energy. His

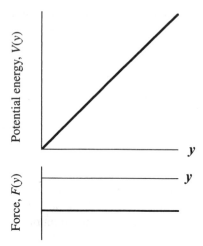

Fig. 6.1 The gravitational potential energy, $V(y)$, of the diver, and the force, $F(y)$, acting on him, plotted as functions of his elevation, y, above the pool level. $V(y) = mgy$, and $F(y) = -\mathrm{d}V/\mathrm{d}y = -mg$. The negative value of $F(y)$ indicates that the force acts in the direction of decreasing y.

velocity is reduced as he enters the water, but his potential energy at pool level is the same as it was before he climbed from pool level to the diving board.

6.4 Force fields

It is often convenient to visualize one-dimensional potential energy by imagining that points equally spaced in potential energy can be extended into horizontal lines in one of the two dimensions not involved, leading in the case of a constant force to a ladder of equally spaced horizontal lines. In two and three dimensions surfaces and closed surfaces would be formed by the process of joining together points of

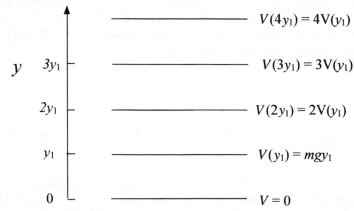

Fig. 6.2 Equipotentials, $V(y)$, of the gravitational force field when the curvature of the earth may be neglected and the acceleration due to gravity, g, may be taken as constant.

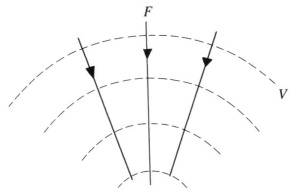

Fig. 6.3 A two-dimensional representation of a few of the equipotentials, V, represented by the circular arcs, and the lines of force, F, represented by the arrowed straight lines, for the gravitational field when the curvature of the earth is not negligible. An equipotential is a line (or surface) on which the potential always has the same value.

equal potential. These imagined energy equipotential lines (or surfaces) define a potential field which extends through space. The gravitational field in our example of the diver would be visualized as a set of parallel, straight lines equally spaced in height above the earth's surface, as shown in Fig. 6.2, because it is a one-dimensional example in which g is constant. If the problem were such that we had to introduce a second dimension, the gravitational equipotentials would be circles centred on the centre of the earth (as shown in Fig. 6.3), while in three dimensions they would form a set of spherical shells.

6.4.1 Conservative fields of force

The diver's potential energy depends only on his height above pool level. At pool level it is the same at the beginning as at the end of his up and down excursion. He has followed a closed one-dimensional path from pool to board and back again and still has the same potential energy. In reality, his path would have involved motion in three dimensions, but even then his potential energy is the same at pool level before and after his dive, whichever route he follows. This independence of potential energy and the total path of the diver as he moves from pool level up and back to pool level is the feature which defines a conservative field of force, in which potential energy depends only on position and remains constant at a particular position (or height in the diver's case). If the diver so misjudged his dive that he landed on the pool side, his potential energy would be still the same as when he started out from the pool side to the board. The sudden loss of energy associated with his coming to a dead stop would be a loss of kinetic energy only. In a conservative field of force the work done in progressing from one point, A, to another, B, is the same as is done in returning back to A, independent of the paths followed between A and B and between B and A. Going from A to B, say, requires that work is done to increase potential energy by $V_B - V_A$. This is work done against the force which is 'minus the gradient of the potential'. In the return from

B to A the same quantity of work is done by the force and, in the case of the diver, is manifested as kinetic energy which may be converted to other forms of energy. The work done against the force is provided from his body mechanisms and the work done by the force is manifested as his kinetic energy. The diver has moved from one point to another in the potential field by one path and has returned to the same point by a different path. In this process he has converted his biological energy into kinetic energy and the only contribution from the potential field has been to provide a mechanism for this energy conversion. When we talk about total energy being the sum of the potential energy and the kinetic energy, we have narrowed down the discussion so that we exclude the mechanism by means of which the potential energy was created in the first place and the mechanism by which the kinetic energy is dissipated afterwards. In a conservative field of force, the sum of kinetic and potential energies is constant; mechanical energy is conserved. There is no conversion of mechanical energy to other forms of energy or vice versa.

There is a clear distinction between the conservative field of force associated with the diver and the forces involved in the inelastic collision problem examined earlier. In inelastic collision the billiard balls lose kinetic energy, which is converted into sound and heat energy when they collide. The event cannot be described completely in terms only of mechanical energy. In non-conservative systems mechanical energy is dissipated into other forms of energy and the forces related to the dissipative effects are called non-conservative forces. Even in 'lossy' systems which lose mechanical energy by conversion into other forms, the sum of the mechanical energy and the other forms of energy is constant. No exceptions to the First Law of Thermodynamics have been found as yet.

In the simple one-dimensional case involving the diver, we have seen that the equipotential lines are parallel to the surface of the earth, while the line of the force of attraction between the diver and the earth is directed perpendicular to the surface of the earth. If the lines of force and the equipotential lines are drawn on the same diagram they are perpendicular to each other. This observation applies also in the two- and three-dimensional cases of free fall towards the earth because radius vectors are perpendicular to tangential vectors. In cases involving more than one dimension, however, there is a slight complication, because force and energy are vector and scalar quantities respectively. This difficulty is overcome quite simply by defining the vector gradient of a scalar quantity as

$$\nabla V = \text{grad}\, V = \mathbf{i}\,\frac{dV}{dx} + \mathbf{j}\,\frac{dV}{dy} + \mathbf{k}\,\frac{dV}{dz},$$

so that the gradient in the statement 'force is minus the gradient of the potential' refers to the vector gradient even in a one-dimensional system.

Example 31. Two bodies of masses $1\,\text{kg}$ and $5\,\text{kg}$ have collinear velocities of $9\,\text{ms}^{-1}$ and $3\,\text{ms}^{-1}$ respectively. The smaller sphere overtakes the larger sphere and collides with it. Assuming that the collision is elastic, what are the velocities of the spheres after the collision?

Before the collision occurs the total momentum of the two body system is

$$p_1 = (1 \times 9) + (5 \times 3) = 24\,\text{Ns}$$

and the kinetic energy of the two bodies is

$$T_1 = 0.5[(1 \times 81) + (5 \times 9)] = 63 \, \text{J}.$$

If the velocities of the bodies after the collision are v_1 and v_2 then, because momentum and energy have been conserved,

$$p_2 = (1 \times v_1) + (5 \times v_2) = 24 \, \text{N s}$$

and

$$2T_2 = (1 \times v_1^2) + (5 \times v_2^2) = 126 \, \text{J},$$

where we have multiplied the kinetic energy equation by two for convenience. From the momentum conservation equation $v_2 = (24 - v_1)/5$. On substituting this value into the energy equation, we obtain

$$v_1^2 - 8v_1 = 9,$$

the solutions of which are $v_1 = 9 \, \text{m s}^{-1}$ or $v_1 = -1 \, \text{m s}^{-1}$. The corresponding values of v_2 are 3 and $5 \, \text{m s}^{-1}$ respectively. The first pair of these solutions are the same as the original values, as might be expected. For this pair of values to be appropriate after the collision process the smaller body must have passed through the larger with neither of them being affected by the event—not a very plausible scenario. The conclusion is that after the collision the lighter body has reversed its direction of travel (indicated by the negative sign) and has a speed of $1 \, \text{m s}^{-1}$, while the larger body continues on its way with an increased speed of $5 \, \text{m s}^{-1}$. The velocity with which the centre of mass of the system moves will be unaltered by the collision, retaining the value $4 \, \text{m s}^{-1}$, because there are no agencies external to the system acting to cause any change in it. In confirmation of the use of relative velocity as a measure of the inelasticity of the collision, notice that the relative velocity in this case has the same value before as after the collision, $6 \, \text{m s}^{-1}$.

Example 32. During a game of pool, after a particular shot, a large number of balls (each of mass $0.5 \, \text{kg}$) are set in motion in a variety of paths around the table with various speeds. Two of these balls are following the same path, with velocities of $5 \, \text{m s}^{-1}$ and $2 \, \text{m s}^{-1}$, the faster one overtaking the slower. If the coefficient of restitution in collision of the balls is 0.75, what is the loss of energy which occurs during the collision? In what directions do the balls travel after they have collided?

Given that the velocities of the balls after collision are v_1 and v_2, conservation of momentum leads to the equation

$$0.5v_1 + 0.5v_2 = 0.5(5 + 2) = 3.5$$

or

$$v_1 + v_2 = 7,$$

while the definition of the coefficient of restitution gives

$$v_1 - v_2 = -0.75(5 - 2) = -2.25,$$

so that

$$v_1 = v_2 - 2.25 = 7 - v_2;$$

in other words,

$$v_2 = 4.625 \, \text{m s}^{-1} \quad \text{and} \quad v_1 = 2.375 \, \text{m s}^{-1}.$$

These velocities have the same sign, so the balls both continue their progress in the same direction as before the collision. The relative velocity of the balls is altered by loss of energy in the collision from $3\,\mathrm{m\,s^{-1}}$ to $-2.25\,\mathrm{m\,s^{-1}}$. The change in sign occurs because before the collision the separation of the balls decreases with time whilst after the collision it increases with time. The energies before and after collision are $0.5[0.5(5^2 + 2^2)] = 29/4$ and $(4.625^2 + 2.375^2)/4 = 27.03/4$, so that the energy loss, the difference between these two, is $0.49\,\mathrm{J}$.

Because no external forces are acting on the system the velocity of its centre of mass is unaltered by the collision, retaining its value of $3.5\,\mathrm{m\,s^{-1}}$. This constancy of the velocity of the centre of mass may be seen as a consequence of the conservation of momentum since $V_c = (p_1 + p_2)/(m_1 + m_2) = P/M$ is unchanged, while the momentum (the sum of the p's) remains constant. The statement that the kinetic energy of the assembly of two (or more) particles observed from outside the assembly, $\frac{1}{2}MV_c^2$, is unchanged, while the kinetic energies of the particles from which it is composed are altered by some form of energy transformation may seem, at first, a little paradoxical. It leads us to distinguish between the energy associated with motion of the centre of mass of an assembly of particles and the internal energy of the assembly, the sum of the energies of its constituent parts. To illustrate this point, let us return to the example of the closed box full of gas, which we examined earlier. The gas molecules are moving about inside the box and the sum of their kinetic energies, the internal energy of the gas–box assembly, is proportional to the temperature of the assembly. If the box has its temperature raised, say by placing it in an oven, the sum of the kinetic energies of the molecules will increase; that is, the internal energy of the assembly is increased. This increase in internal energy will not cause the box to move about. Its centre of mass will remain fixed in place, the velocity and of the centre of mass and the kinetic energy associated with the centre of mass will be unchanged. The internal energy of the assembly is not available to do the work necessary to change the velocity of the centre of mass. This unavailability of energy is most often expressed in terms of the Second Law of Thermodynamics, which can be seen as a statement that it is impossible to make a perpetual motion machine, even one more general than the one we cited before.

Example 33. The interactions between atomic particles and the nuclei with which they collide have been the subject of many experiments. Such experiments provide information about the energy exchanges which take place when the particles collide with the nuclei and produce excited states of the original nucleus or transmute it to another species. A symbolic way of representing a reaction of this type, in which a particle x is incident on a nucleus X, producing a nucleus Y and a particle y, is as $X(x, y)Y$. X and Y may be different species or the same species of nucleus, while x and y may be the same particle or different particles. In such a reaction the difference between the sum of the kinetic energies of x and X and the sum of the kinetic energies of y and Y is called the energy balance of the reaction, Q. In reactions of this type, the masses of X and Y are often much greater than are those of x and y, so that the kinetic energy of the Y product is small and difficult to measure. Making the assumption that simple collision

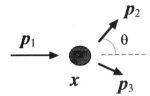

Fig. 6.4 A schematic momentum diagram for the reaction $X(x, y)Y$.

mechanics can be applied to a nuclear interaction, and that the target nucleus X is a nucleus in a solid target material, effectively at rest, find an expression for Q which does not involve the velocity of the product nucleus Y.

Because the kinetic energy of X, $T_X = 0$, the equation for the energy balance reduces to

$$Q = T_Y + T_y - T_x = T_3 + T_2 - T_1,$$

where the substitution of number for letter subscripts is made to reduce the possibility of confusion. Since there is no reason to assume that the product particles travel along the same line, the momenta of the particles before and after collision will be as shown schematically in Fig. 6.4. In this two-dimensional problem it is convenient to use a vector representation of the momenta p_3, p_2, and p_1 which are those of the Y, y, and x particles respectively. Remembering that momentum and kinetic energy are related by

$$T = p^2/2m$$

we may write three relations of the form

$$T_i = p_i^2/2m_i \qquad \text{for } i = 1, 2, \text{ and } 3,$$

where the subscripts 1, 2, and 3 refer to x, y, and Y respectively. We want to eliminate T_3 from the Q-equation and can do this because, referring to Fig. 6.1, it is seen that

$$p_3 = p_1 - p_2$$

so that

$$p_3^2 = |p_3|^2 = p_1^2 + p_2^2 - 2p_1 \cdot p_2,$$

giving

$$T_3 = \frac{p_1^2 + p_2^2 - 2p_1 p_2 \cos \theta}{2m_3}.$$

When this expression is substituted into the Q-equation and the kinetic energies substituted in place of the momenta, we obtain

$$Q = [1 + (m_2/m_3)]T_2 - [1 + (m_1/m_3)]T_1 - \frac{2[(m_1 m_2 T_1 T_2)]^{1/2}}{m_3} \cos \theta.$$

The ability to write the energy balance equation in this form means that it is not necessary to measure T_3 in order to find Q.

Exercises

6.1. A ball is dropped from a height of $1\,\mathrm{m}$ on to the ground beneath. If the coefficient of restitution for impact between the ball and the floor is 0.9, what time will elapse before the ball comes to rest?

6.2. A fast bowler bowls a ball of good length which has a velocity of $40\,\mathrm{m\,s^{-1}}$ when it strikes the pitch. If this velocity is directed at an angle of $30°$ to the plane of the pitch and the coefficient of restitution is 0.6, what is the velocity of the ball after it bounces, and in which direction is it travelling?

6.3. In a snooker game, a coloured ball and the cue ball lie on a line $20\,\mathrm{cm}$ from, and parallel to, one of the sides of the table, with the colour positioned so that it is $40\,\mathrm{cm}$ away from a corner pocket. If the cue ball is well struck so that the colour is 'potted', what is the angle through which the path of the cue ball is deflected by the collision if it is an elastic collision?

Solutions

6.1. The force acting on the ball as it falls is its own weight so that its acceleration is g and the time it takes to reach the ground from a height h is

$$t_0 = \sqrt{(2h/g)}$$

and its velocity on impact is $\sqrt{(2gh)}$. After the first impact the velocity of the ball will be reduced by the factor e, the coefficient of restitution, so that $v_1 = 2gh$. It will lose this upward velocity a time given by

$$t_1 = 2v_1/g = e\sqrt{(2h/g)}$$

and its return to the ground will also take time t_1. The velocity after the second bounce will be $v_2 = e^2\sqrt{(2h/g)}$, and so on. The total time taken in bouncing is thus

$$T = t_0 + t_1 + t_2 + t_3 + \cdots$$
$$= \left[\sqrt{(2h/g)}\right][1 + 2(e + e^2 + e^3 + \cdots)].$$

The series in e is a geometrical progression and, since it converges (because $e < 1$), can be summed even if it has an infinite number of terms. The ball comes to rest, in principle at least, after an infinite number of bounces and the series has to be summed for an infinite number of terms. The sum to infinity of $(1 + e + e^2 + \cdots)$ is $[1/(1 + e)]$, so that

$$T = \left[\sqrt{(2h/g)}\right][1 + 2e(1/(1 - e))]$$
$$= \frac{1 + e}{1 - e}\sqrt{(2h/g)}$$

with the consequence that, for $h = 1$, $e = 0.9$, and $g = 9.81$, we have $T = 8.58\,\mathrm{s}$. If e had the value 0.8 the time to come to rest would be $4.06\,\mathrm{s}$, while for

$e = 0.5$ it would be only 1.35 s. These times include the time for the initial fall and the times measured from the time of the first impact on the floor would be 8.13, 3.61, and 0.9 s for $e = 0.9$, 0.8, and 0.5 respectively. This very marked dependence of T on e emphasizes that the energy loss on impact is determined by $(1 - e^2)$. When $e = 0.9$ the ball loses 19% of its energy at the first impact, for $e = 0.8$ it loses 36%, and for $e = 0.5$ it loses 75% of its energy on first impact.

6.2. The components of the velocity are $40 \sin 30° = 20$ perpendicular to the pitch and $40 \cos 30° = 34.6$ parallel to it. The parallel component is unaltered by the impact, while the perpendicular component is reversed in direction and reduced by the factor e, the coefficient of restitution. The velocity components after the bounce are $20 \times 0.6 = 12$ and 34.6, so that the velocity after the bounce is 36.7 m s^{-1}. The angle between the direction of this velocity and the surface of the pitch is $\tan^{-1}(12/34.6) = 19°$.

   If the coefficient of restitution were 0.8 the velocity and angle would be increased to 38.2 m s^{-1} and 24.8°. This makes it obvious why fast bowlers prefer new balls to old ones, and hard, dry pitches to softer, rain-affected ones, and why batsman dislike pitches with 'uneven bounce'. The ball doesn't have to be 'stuck in the mud' to lose a considerable proportion of its velocity in the bounce. On the other hand, treating a cricket ball as a point mass as we have done is a gross oversimplification. If the ball were simply spherical the spin imparted to it by the bowler and by its impact with the pitch would have a marked effect on its bounce. In addition, the ball has a seam around its centre which changes its rotational symmetry from that of a sphere, which is symmetrical under the operation of any rotation about any diameter, to a state in which this complete rotational symmetry occurs by rotation only about the diameter perpendicular to the plane of the seam. The orientation of the seam and the way in which the ball is spinning, as well as the exact shape of the pitch at the point of impact, will all affect the bounce of the ball.

6.3. In this collision the momentum vector triangle is described by $m_1 v = m_1 v_1 + m_2 v_2$, where the m's are the masses of the balls, v the initial velocity of the cue ball, v_1 its velocity after impact, and v_2 the velocity of the colour after the collision. Since the balls have the same mass this is simply $v = v_1 + v_2$. The collision is elastic and kinetic energy is conserved so that, cancelling out the factor $\frac{1}{2}m$ in the energy equation, we may equate the kinetic energy and the square of the momentum (divided by the mass) conservation equation as

$$v^2 = v_1^2 + v_2^2 = |v|^2 = |v_1|^2 + |v_2|^2 - 2v_1 \cdot v_2.$$

This equation is satisfied only if $v_1 \cdot v_2 = 0$; that is, the angle between the directions of v_1 and v_2 is 90°. For the colour to enter the pocket after the impact its velocity must be directed towards the pocket so that the angle between the original path of the cue ball and the direction of v_2 is 150°. The supplement of that angle is 30° so that the path of the cue ball is deflected by 60°.

   In actuality, the problem is complicated because the balls are not point masses, and the collision is not elastic but has a coefficient of restitution, e.

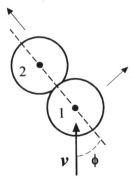

Fig. 6.5 A two-dimensional representation of the snooker ball collision. The cue ball is 1 and the colour is 2.

In this case the meaning of the term 'relative velocity' needs to be clarified. A two-dimensional representation of the moment of impact is shown in Fig. 6.5, in which the line passing through the centres of the discs defines a line of centres. The impact takes place along the line of centres so that the impulse that occurs is directed along the line of centres. Any change in momentum is a change of momentum along that line—there is no impulse perpendicular to the line of centres; that is, the component of v perpendicular to the line of centres is not altered by the ball collision. The energy associated with that velocity component is unchanged and will not have a coefficient of restitution associated with it. The relative velocity to be used in the expression for the energy loss is thus the relative velocity along the line of centres. This means, for the situation shown in Fig. 6.5, remembering that the balls have equal masses, that

$$v_1 - v_2 = -ev \cos \phi$$

where v_1 and v_2 are the components of velocity after collision of the balls along the line of centres. Conservation of momentum leads to the equation

$$v_1 + v_2 = v \cos \phi,$$

so that

$$v_1 = \frac{(1-e)}{2} v \cos \phi \quad \text{and} \quad v_2 = \frac{(1+e)}{2} v \cos \phi.$$

The direction of motion, θ, of the cue ball relative to the line of centres after collision is given by

$$\tan \theta = \frac{v \sin \phi}{\frac{1}{2}(1-e)v \cos \phi} = \frac{2 \tan \phi}{(1-e)},$$

the ratio of its velocity components perpendicular to and along the line of centres. For elastic collision $e = 1$ so that θ is always 90°. In the case when $e = 0.8$, simple substitution leads to $\theta = 80°$. Such a substitution would be inadequate to deal with the case in which the balls lie initially along a line parallel to the side of the table, since the cue ball would be aimed to make contact along the appropriate line of centres, so that ϕ would be more than 30°. The ball diameter would then be required to solve this problem.

7
Tensile forces and projectiles

7.1 Tensile forces

We have seen that two collinear forces that are equal in magnitude and opposite in direction, acting on the same body, nullify each other so that the body is not subjected to any acceleration. Imagine the situation in which a rope is attached at one end to a well built wall. When we pull on the free end of the rope the rope straightens out but nothing else happens, provided that the rope does not stretch unduly. A force F is exerted on the end of the rope away from the wall and this is nullified by the force, $-F$, exerted by the wall. Since the force exerted on and by the wall is transmitted along the intervening rope, the rope itself must exert a force on the wall which is equal to the force being exerted on the rope there. At the end of the rope attached to the wall there is a force F exerted by the rope. A second force, $-F$, at the other end of the rope must be exerted by the rope to nullify the pulling force, F. These two forces in the rope are known as the tension in the rope. The statement that the tension in the rope is F means that the rope is subjected to tensile forces which, in effect, have a magnitude F at either end of the rope. The tension in the rope is the means by which the rope transmits force from one of its ends to the other. A rod, or any solid material, could be substituted for the rope as the agent for transmitting the force. This rod, or solid, would be subjected to tension (as shown in Fig. 7.1) during its 'transmission' of the force along its length. Unlike the rope, the rod or solid could be pushed, rather than pulled, at its free end, in which case the forces in the material would be directed out from its bulk instead of into its bulk. The forces acting on the solid would tend to compress it and the reaction from the solid would be directed in opposition to the applied forces and the rod or solid would be 'in compression'.

So far, the rope, or rod, has been taken as rigid, which means that it will neither increase nor decrease its length under the influence of tensile or compressive forces respectively. What happens when the rod isn't rigid, when it has elastic properties, or if it stretches when subjected to tensile forces? Just as in the case of the rigid rod a force F exerted on the free end of an elastic rod attached by one

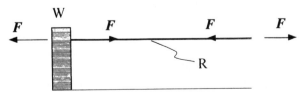

Fig. 7.1 The forces acting when a rod, R, is attached to a wall, W, and subjected to a tensile force, F, at its free end. The forces shown acting in the rod represent the tension in the rod.

end to a wall will introduce a tension into the elastic rod (which could be, for example, a spring) and the tension will increase until it is equal in magnitude to F, at which stage F would be nullified. The rod will then be subject to a tension F which will transmit the force F to the wall. The only difference between the cases of the rigid rod and the elastic one is that the elastic rod has the capacity to stretch under the influence of tensile forces exerted on its ends. For the rigid rod the force exerted on its end is applied always at the same point, its point of application does not move, and no work is done in the process of exerting force on the wall. The opposite is true for the elastic rod—work has to be done by the force as its point of application moves while the rod is stretching. This work, done by the force, has to be manifested as energy and, since the system remains effectively at rest, it has to be a potential energy. It is a potential energy stored in the elastic rod for if the force is removed from the rod it will revert to its original length.

Say that the force F applied to the elastic rod is a force in the x-direction which causes an extension δx in the length of the rod. The work done is

$$F\delta x = W = -\delta V,$$

where V is the potential energy of the rod. The force required to increase the length of an elastic rod by x is established from experiment as

$$F = -kx,$$

where k is the spring constant of the rod. This linear relation between the force acting on an elastic rod and the extension it produces is known as Hooke's Law. The tension in the extended rod is

$$-F = kx,$$

so that if the force F is removed from the end of the extended rod there is an 'internal' force $-kx$ (a restoring force because of the negative sign) acting on it to make it return to its initial length. The work done by the force in producing the extension x in the length of the rod is

$$W = -\int_0^x kx\,dx = -\tfrac{1}{2}kx^2$$

so that

$$V = \tfrac{1}{2}kx^2.$$

This increase in the potential energy of the rod tells us that the rod will revert to its original length when the external force is removed. When the rod is unstretched its potential energy will be a minimum ($x = 0$) and if we identify this minimum energy as the zero of potential energy we may say that V, the change in potential energy, is just $V(x)$, the potential energy associated with the extension of the rod; that is, $V(x) = \tfrac{1}{2}kx^2$. The parabolic form of the potential energy of the rod indicates that the rod will revert to its original length when the force is removed, whether the force is compressive or tensile, because the minimum potential energy occurs for $x = 0$, corresponding to the natural length of the rod.

7.2 Electric and gravitational potential

The use of the word 'potential' in the discussion of electrical problems is an indication that the concepts of electrostatic potential and electrostatic field have their origin in energy considerations. The basic specification of these comes from Coulomb's experimentally established law of electrostatic attraction or repulsion. The force acting between two point charges, q_1 and q_2 coulombs, separated by a distance r in the direction defined by the unit vector u_r, is given by

$$F = \frac{q_1 q_2}{4\pi\epsilon_0 r^2} u_r$$

where $4\pi\epsilon_0 = 1.1 \times 10^{-11}$ involves the permittivity, ϵ_0, of free space. This form of the equation for the force suggests that we could describe the two-charge situation by a potential energy field and, for example, if r lies along the x-axis and the separation of the charges is taken as x then

$$\int F \cdot dr = \frac{q_1 q_2}{4\pi\epsilon_0} \int \frac{dr}{r^2} = -\frac{q_1 q_2}{4\pi\epsilon_0 x} = -V(x).$$

This expression is changed from potential energy to electrostatic potential simply by setting $q_2 = 1$ and $q_2 = q$, giving the electrostatic potential energy surrounding a point charge q as

$$V_E(r) = (q/4\pi\epsilon_0 r) \quad \text{volts},$$

where we are allowed to revert back from the one-dimensional parameter x to r because of the spherical symmetry associated with a single point charge. The electrostatic field is then simply $E = -\nabla V_E = -u_r(dV_E/dr) = u_r(q/4\pi\epsilon_0 r^2)$ volts m^{-1} which will produce a force of $-\nabla V_E$ on each unit of charge situated at a distance r away from the charge q. These considerations lead to the well known observations that the potential energy of a charge q situated at a point where the electrostatic potential is V_E is

$$V = qV_E$$

while the force acting on the charge is given by

$$F = qE.$$

Obviously, if we can calculate the electrostatic potential surrounding a particular distribution of charge we will be able to derive the form of the electric field and predict the form of the field that would act on a charge situated at any point in the field. The advantage in dealing with electrostatic potential rather than electrostatic field is that the potential (derived from energy considerations) is a scalar quantity, while the field (derived from considerations of force) is a vector quantity.

This electrostatic potential is a mathematical analogue of the gravitational potential which may be derived from the force law for gravitational attraction;

namely, that the force acting between two bodies of masses M and m separated by a distance r is given by

$$\text{gravitational force} = -\frac{GMm}{r^2}\,\boldsymbol{u}_r$$

while

$$\text{gravitational potential} = -(GMm/r),$$

where G is the gravitational constant (the minus signs indicate attractive force and potential). In cases in which we can treat the acceleration due to gravity, g, as constant—that is, close to the surface of the earth, we may say that $g = (GM/R^2)$, where R is the radius of the earth. The potential energies in the electrostatic field and in the gravitational field depend only on position, so that both of these may be considered to be conservative fields of force.

7.3 Bound and free states

Example 34. A planet circulating in its orbit around the sun is subjected to a force of the form $F = -K/r^2$, where K is a positive constant and r is the separation of the planet from the sun. The planet Venus has an orbit which is practically circular, with a radius of 1.08×10^{11} m. If the mass of the planet is 4.9×10^{24} kg and its 'year' is 1.94×10^7 s, what is the total energy of the planet?

It might appear at first that insufficient information has been given, but this is not the case. The force has the form $F(r) = -K/r^2$, so that the potential has the form $V(r) = -k/r$, and in order to progress further we need to known the value of K. Since we have a body (the planet) travelling in a stable circular orbit (in other words, it is not moving towards or away from the sun), there is no force acting on it to alter the orbit. This means that the attractive gravitational force must be nullified by the centrifugal force which is required to change the direction of the planet's velocity. Since this has the form mv^2/r, we may write

$$F(r) = K/r^2 = mv^2/r$$

or, since the planet's kinetic energy is $T = \frac{1}{2}mv^2$, multiplying F by r,

$$K = mv^2r = 2Tr, \quad \text{giving} \quad V(r) = -2T.$$

The potential energy is negative and has a magnitude twice that of the kinetic energy of the planet. The total energy is the sum of the kinetic and potential energies; that is,

$$H = T + (-2T) = -T,$$

so that with $T = \frac{1}{2}mv^2$ the total energy of Venus in its orbit is

$$H = -T = -[m/2][2\pi r/t]^2$$

(where t is the time for one orbital revolution, the 'year')

$$= -3 \times 10^{33} \text{ J}, \quad \text{or} \quad -3 \times 10^{27} \text{ MJ}.$$

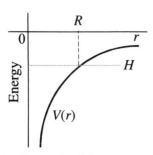
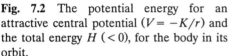

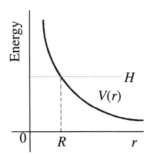

Fig. 7.2 The potential energy for an attractive central potential ($V = -K/r$) and the total energy H (< 0), for the body in its orbit.

Fig. 7.3 The potential energy for a repulsive central potential ($V = K/r$) and the total energy H of a body in that potential field.

The total energy of the planet is negative! To examine the significance of this result we draw a graph of potential energy against r as shown in Fig. 7.2, in which the curve V ($= -K/r$) is the attractive potential in which the body (the planet) moves, and H, the total energy, is negative. The gravitational field is a conservative field of force so that

$$H = T - K/r$$

or

$$T = H + K/r$$

and the body gains energy as r decreases. If $H < 0$, as shown in Fig. 7.2, then for one value of the separation R of the body from the centre of attraction we have $T = [-H + (K/R)] = 0$, so that the body has zero kinetic energy for that value of r. For values of $r < R$ the kinetic energy $T > 0$, while if $r > R$ the body would have $T < 0$. The negative kinetic energy implied by $T < 0$ is an impossible situation, since the quantities involved in T, m and v^2, are both positive. The conclusion reached is that the body is confined within a sphere of radius R centred on the centre of attraction. The body is bound to the centre of attraction—it cannot escape from the sphere of radius R unless it is provided with sufficient energy to make its total energy positive. When $H > 0$ the line representing H never crosses the line representing $V(r)$ and the kinetic energy of the body, $T = H + K/r$, is greater than 0 for every value of r. A state with negative total energy is a bound state of the system which requires an infusion of energy for it to be broken up. A state of positive energy is a 'free' state of the system in which the body never enters into a stable bound state.

If the potential were repulsive, say of the form K/r, then, as shown in Fig. 7.3, both H and V would always be positive. Since we may write

$$T = H - K/r,$$

T will be zero when r has the value R defined by $H = K/R$ and $T < 0$ for $r < R$. The body cannot penetrate into a sphere of radius R centred at the centre of repulsion because inside the sphere its kinetic energy would have to be negative. In this case R defines a distance of closest approach of the body to the centre of repulsion. There is no possibility of forming a bound state when the potential is repulsive, since that would require $T < 0$.

Although we have taken a case of gravitational attraction as example, we could have used as a model the case of negatively charged electron travelling in a closed orbit about a positively charged nucleus. This is the sort of model that was studied by Lord Kelvin and J. J. Thomson when they tried to provide explanations of the structure of the atom. Their models did not prove satisfactory because an accelerated electron emits electromagnetic radiation (a non-conservative effect) and a many-electron atom would suffer from instabilities analogous to those expected in Saturn's rings (before Maxwell solved that particular problem). The problem of the stability of the atom was not resolved until Bohr adapted mechanics by the addition of quantum hypotheses, which are not our concern here.

Example 35. Kitchen scales used for weighing ingredients for cooking depend usually on the compression of a spring for their operation. If in such a set of scales a weight of 20 N compresses the spring by 1 cm, what weight will compress it by 2 cm? How much work will be done in this second compression?

The compressive stress in the spring compressed by 1 cm has the same magnitude as the force exerted to produce the compression, $F = kx$; that is, $20 = k \times 10^{-2}$, or $k = 2\,\text{kN}\,\text{m}^{-1}$. When $x = 0.02\,\text{m}$, $F = 40\,\text{N}$. A further 20 N is required to double the compression of the spring, as would be expected from the linear relationship between force and compression.

The whole work done in compression may be evaluated by either of two methods. The first is to find the product of force and distance moved in the direction of the force. To use this method we need to know the force acting on the end of the spring (the point of application of the external force) during the compression. If the compression at any time between the initial application of the force and the attainment of equilibrium is s, then the force $F(s)$ acting at the end of the spring is a combination of the applied force F_0 and the opposing force, $-ks$, due to the compression of the spring; that is, $F = F_0 - ks$. The work done by this force in compressing the spring to the equilibrium position, x, is

$$W = \int_0^x F(s)\,ds = \int_0^x (F_0 - ks)\,ds = F_0 x - \tfrac{1}{2}kx^2.$$

Since for the equilibrium state $F_0 = kx$, $W = \tfrac{1}{2}kx^2$, meaning that the work done in compressing the spring from 1 cm to 2 cm is 0.3 J. The point raised here is that the force in the force–distance product is not constant, but varies with distance, from F_0 for $s = 0$ to zero at $s = x$. Had we taken a naive approach by assuming the force to be constant, we would have overestimated the work done by a factor of two.

The second approach is to identify the work done with the change in potential energy of the spring, $W = \delta V$. Since $V = \tfrac{1}{2}kx^2$, the work done is 0.3 J as obtained already.

Example 36. A diver projects himself from a diving board and falls to the pool 6 m below the board. What will be his velocity when he enters the water?

The potential energy of the diver at the height, z, of the board (measured from some arbitrary datum) is mgz, where g is the acceleration due to gravity. His potential energy at pool level is mgy, where y is the height of the pool surface above the datum. In falling to pool level the change in the diver's potential energy is

$$\delta V = mg(z - y) = mgh,$$

where h is the height of the board above pool level. The gravitational field of force is conservative and so this change in potential energy is equal to the change in kinetic energy; that is, since his initial velocity is zero his velocity at pool level is obtained from

$$\tfrac{1}{2}mv^2 = mgh \qquad \text{or} \qquad v = \sqrt{(2gh)}$$

leading to $v = 10.85\,\mathrm{m\,s^{-1}}$.

In a detailed consideration of the diver's fall, the resistance to his motion by the air through which he falls would be included in the calculation. We have ignored this non-conservative force, because its effects are relatively small when the height of fall is small and the diver's velocity is not very large. Had the diver been a sky diver falling from a height of 1 km we could not have neglected the effect of air resistance on his motion which—as we estimated in Example 25—is probably proportional to the square of his velocity.

Example 37. Two springs AB and CD of equal unstretched length, l, with spring constants k_A and k_C respectively, are to be used to nullify a force of 10 N which acts parallel to the smooth surface on which the springs rest and perpendicular to the wall situated at one of their ends. The compression produced by the force acting on the spring AB alone is 2 cm and the corresponding quantity for CD is 3 cm. Neither 2 nor 3 cm is a small enough compression to be satisfactory, so that the possibility of applying the force to both springs at the same time has to be considered. The springs may be connected in series, with A at the wall and B and C connected so that the unstretched length of the combination spring is $l_s = 2l$, or they may be connected in parallel when A and C are situated at the wall and the force acts on B and D together, in which case the unstretched length of the combination is $l_p = l$. Which combination should be used to make the compression smaller?

The same force, say F, acting on the individual springs produces a compression of x_A and x_C, respectively, so that the spring constants are $k_A = 10/0.02 = 500\,\mathrm{N\,m^{-1}}$ and $k_C = 333\,\mathrm{N\,m^{-1}}$. When the two springs are arranged in series the same tension exists throughout their combined length, so that if the compression is then X_s and their 'combined spring constant' is K_s, then

$$F = K_s X_s.$$

The compression of the combined spring is made up of a compression x_1 of AB and a compression x_2 of CD, so that we could just as well have written

$$F = K_s(x_1 + x_2),$$

where, of course, $x_1 = F/k_A$ and $x_2 = F/k_C$. On substituting these values and cancelling F, we obtain

$$\frac{1}{K_s} = \frac{1}{k_A} + \frac{1}{k_C},$$

which indicates that the spring constant of a series combination of springs is always less than the spring constants of the springs from which it has been formed. In this case it would be $200\,\mathrm{N\,m^{-1}}$, so that the new compression would be $5\,\mathrm{cm}$, larger than the compression of either of the individual springs.

When the two springs are arranged in parallel, the equilibrium state will occur when their compressions are equal, say X_p. Then

$$F = K_p X_p = k_A X_p + k_B X_p,$$

leading to

$$K_p = k_A + k_C,$$

showing that the spring constant of a parallel combination of springs is always greater than any of the spring constants of its component springs. In this case K_p would be $833\,\mathrm{N\,m^{-1}}$ and X_p would be $1.2\,\mathrm{cm}$. The smaller compression would occur when the springs were arranged in parallel.

It's of interest to notice that the formulas for the spring combinations are analogues of the formulas for combining capacitors in parallel or in series. The analogy might be made as well between combinations of springs and adding thermal conductivities in series and parallel or adding conductances (reciprocal resistances) in series or in parallel. These analogies are simple mnemonics rather than exact similarities.

Example 38. A burglar always carries to his work a $20\,\mathrm{m}$ length of nylon rope so that, if necessary, he can make a rapid exit through a window well off the ground. During one expedition he fastens his rope to a pulley support above a warehouse window the sill of which is $15\,\mathrm{m}$ above the ground. Being disturbed, he runs to the window, climbs on the sill, and reverses his position so that he faces the wall. About to leave the sill, he notices a sign at the side of the pulley, 'Maximum Load $60\,\mathrm{kg}$'. If the rope has a mass of $10\,\mathrm{kg}$, the burglar has a mass of $70\,\mathrm{kg}$, and he launches himself off the sill to avoid capture, what will be his minimum velocity when he reaches the ground?

The rope weighs $98\,\mathrm{N}$ and the pulley support will break if subjected to a force of more than $588\,\mathrm{N}$. The 'space capacity' of the support is the difference between these, $490\,\mathrm{N}$. The burglar's weight is $686\,\mathrm{N}$. If he simply hangs on to the rope its support will break and he will fall to the ground with an acceleration of $g = 9.8\,\mathrm{m\,s^{-2}}$. To reduce the tension in the rope so that it is less than $490\,\mathrm{N}$ he will have to allow himself to accelerate downwards (by loosening his grip on the rope) with an acceleration of $a\,\mathrm{m\,s^{-2}}$. The tension in the rope, the force he exerts on the rope, will be $70(9.8 - a)$. This force must be less than $490\,\mathrm{N}$, giving $70(9.8 - a) < 490$ if the support is not to break; that is, $a > 2.8\,\mathrm{m\,s^{-2}}$. The minimum velocity with which the burglar will reach the ground is $v = (2ah)^{1/2} = (30 \times 2.8)^{1/2} = 9.7\,\text{m s}^{-1}$, approximately half the maximum speed possible (when the support breaks as he leaves the window sill).

Alternatively, the forces acting on the burglar are his weight and the tension in the rope. His equation of motion is $mg - \mathbf{T} = ma$, or $\mathbf{T} = m(g - a)$. Provided that $\mathbf{T} < 490\,\text{N}$, the support will remain intact.

Rounding g to $9.8\,\text{m s}^{-2}$ can be justified by observing that it introduces an error of 0.1%, which is much less than the daily variation in the burglar's mass.

7.4 Projectiles

Example 39. An anti-aircraft gun has its barrel pointing vertically upwards in the hope of shooting down an aircraft passing directly overhead. If the velocity of a shell leaving the barrel is $300\,\text{m s}^{-1}$ and the aircraft flies at an altitude of 4500 m, is the hope of a successful 'kill' ill-founded?

Taking the zero of height to be the upper end of the gun barrel, we may say that the energy of the shell as it leaves the barrel is $T = \frac{1}{2}mv^2 = 45\,\text{kJ}$. In the conservative gravitational field of force the shell will reach its maximum altitude, h, when this kinetic energy has been converted all to potential energy, V, with $V = mgh$. Hence $h = 4592\,\text{m}$. The maximum height attainable by the shell is such that, if it were aimed correctly, it could strike the aircraft.

Alternatively, the velocity of the shell would be zero when it had attained its maximum altitude. The downward acceleration of the shell would be g and the equation

$$v_2^2 = v_1^2 - 2gh$$

with $v_2 = 0$ leads to the same value of the maximum altitude of the shell.

Here we have neglected the effect of dissipative forces, such as resistance by the air to the motion of the shell. In reality, the existence of these forces would lead to work being done by the shell to overcome them. Some of the initial kinetic energy would be used up in doing that work and the highest point of the shell's trajectory would have an elevation considerably less than the 4592 m obtained when dissipative forces were neglected. The aircraft would pass by unscathed.

Example 40. The elevation angle of a cannon is adjusted so that its barrel is inclined at an angle of 30° to the horizontal. When the cannon is fired the shot leaves its barrel with a velocity of $300\,\text{m s}^{-1}$. What is the form of the subsequent path of the shot? What will be the maximum altitude attained by the shot? What will be the range of the shot if the cannon stands on a horizontal plane?

This question is different from Example 39 only because the cannon is not aimed vertically. In the absence of resistive forces the shot, after it leaves the barrel, is subjected to one force only, the weight of the shot acting vertically downwards. If we take the position of the cannon as the origin of a coordinate system, the horizontal displacement of the shot from the cannon as x, and the corresponding vertical displacement as y, then there is no force acting in the x-direction and a force $(-mg)$ acting in the y-direction. In the absence of an x-directed force the x component of the initial velocity, v_x, will remain constant so that the displacement in the x-direction at a time t after firing the cannon will be

$$x = v_x t = vt \cos 30°.$$

In the y-direction the equation of motion will be

$$m \frac{d^2 y}{dt^2} = -mg.$$

On integration this leads to

$$y = v_y t - \tfrac{1}{2} g t^2,$$

where y is the vertical displacement of the shell and $v_y = v \sin 30°$ is the y-component of the initial velocity of the shell. Here we have two equations, one for x and one for y, parametric in time t. If we eliminate t from them we will obtain an equation relating x and y directly. From the x equation we may set $t = x/v_x$ and, on substituting into the y equation, we obtain

$$y = x \tan 30° - \frac{g x^2}{2 v^2 \cos^2 30°},$$

which is the equation of a parabola. The maximum height attained by the shell occurs when the y-component of the shell's velocity is zero,

$$v_y(t) = v_y - gt \quad \text{or} \quad v_y = gx/v_x;$$

that is, it occurs for $x = (v^2/g) \cos 30° \sin 30°$ when $y = h = (v^2/2g) \sin^2 30°$. The range, R, of the shot is the value of x for which $y = 0$; that is, neglecting the case $x = 0$, which is the initial condition,

$$R = (2v^2/g) \cos 30° \sin 30° = (v^2/g) \sin 60°.$$

The equation describing the trajectory of the shot is

$$y = \frac{x}{\sqrt{3}} - \frac{g x^2}{3 v^2},$$

the maximum height attained is $h = v^2/8g$, and the range of the shot is

$$R = \frac{v^2 \sqrt{3}}{2g}.$$

What is noticeable about these results is that the maximum value of y occurs at a position $x = \tfrac{1}{2} R$.

This method of considering velocity components in isolation has been used to examine the motion in two dimensions of a body subjected to a one-dimensional force. It is a quite general method which may be used to model, for example, a charged particle moving into a uniform electric field, as well as a projectile moving in a gravitational field.

The solution of Example 40 is generalized simply by substituting the angle θ, which is an angle in the range $0 < \theta < \frac{1}{2}\pi$, for the 30° used in the example. Then the form of the trajectory becomes

$$y = x \tan \theta - \frac{gx^2}{2v^2 \cos^2\theta},$$

the range $R = (v^2/g)\sin 2\theta$, and the maximum height attained is $h = (v^2/2g)\sin^2\theta$ at $x = \frac{1}{2}R$. The maximum range, R_m, of the projectile corresponds to the maximum value of $\sin 2\theta$, $\theta = \frac{1}{4}\pi = 45°$, with $R_m = v^2/g$. The maximum value of h corresponds to $\sin^2\theta = 1$, $\theta = \frac{1}{2}\pi$, and is attained when the projectile is fired vertically upwards; $h_m = v^2/2g$. As long as neither h nor R is zero $(\theta \neq 0, \theta \neq 90°)$ the equation to the trajectory may be written in a different form by noticing that $h = (R \tan \theta)/4$ and $(g/2v^2 \cos \theta) = (\sin \theta)/R$. This leads to

$$y = \frac{4h}{R}x\left[1 - \frac{x}{R}\right],$$

an equation which does not involve θ explicitly.

Example 41. A small boy at the top of a vertical cliff 20 m high attempts to throw a pebble into the sea below. He throws the stone at $20\,\mathrm{m\,s^{-1}}$ at an angle inclined 30° above the horizontal. If the sea is separated from the base of the cliff by a 50 m wide rock formation, will he succeed in making a splash?

Relative to the boy the vertical coordinate describing sea level is $-20\,\mathrm{m}$, and the vertical component of the initial velocity of the pebble is $20 \sin 30° = 10\,\mathrm{m\,s^{-1}}$. Since

$$y = v_y t - \tfrac{1}{2}gt^2$$

we have, taking $g = 10\,\mathrm{m\,s^{-2}}$ so that the coefficients in the equation are integers,

$$-20 = 10t - 5t^2,$$

an equation that is quadratic in t. The solution of this equation is $t = 3.24\,\mathrm{s}$. In this case the pebble will travel a horizontal distance $x = v_x t = vt \cos 30° = 56.05\,\mathrm{m}$. The pebble will fall into the sea.

We might have written the parabolic equation for the trajectory in terms of x and y. Its solution is $x = 56.05\,\mathrm{m}$, as might be expected. Taking $g = 9.81\,\mathrm{m\,s^{-2}}$ leads to $t = 3.281\,\mathrm{s}$ and $x = 56.83\,\mathrm{m}$. A 2% error in g has led here to a 1.2% error in x.

So far we have examined the behaviour of a projectile fired from a position on a horizontal plane. Such a plane, flat and level within the range of the projectile, is perhaps inadequate to represent actual situations. It's far more likely that the launch and landing points for the projectile are separated in elevation by some height difference, h. A line joining the two points would be inclined, say, at an angle α to the horizontal. How does the existence of this slope alter the maximum range of the projectile and the optimum angle of projection?

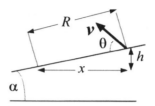

Fig. 7.4 A projectile projected at an angle θ with respect to a plane inclined at an angle α with respect to the horizontal.

Take as an example the situation shown in Fig. 7.4, in which a projectile is to be fired from a point on a slope inclined at an angle α to the horizontal. Let the angle of projection measured relative to the plane be θ, and say that the range of the projectile on the plane is R. The displacements of the launch and landing points are then h vertically and x horizontally, as shown in Fig. 7.4. We use the same principles we used in the case of the horizontal plane but follow a slightly different route. First of all let the normal component of the projectile velocity, v, be the component of the velocity normal to the inclined plane, and use the subscript z to identify this component. Then $v_z = v \sin \theta$ and the acceleration suffered by the projectile in the z-direction (normal to the plane) is the component of g normal to the plane, $g \cos \alpha$. The displacement z, normal to the plane, at time t after projection, is

$$z = vt \sin \theta - \tfrac{1}{2}gt^2 \cos \alpha,$$

so that

$$z = 0 \qquad \text{for } t = 0 \quad \text{and for} \quad t = \frac{2v \sin \theta}{g \cos \alpha},$$

which statements define the time of flight of the projectile. Then, because it makes for simpler algebra, we take the component of the velocity which is horizontal, v_x (which is not parallel to the plane) and multiply it by the time of flight to obtain the horizontal displacement of the landing point from the launch point:

$$x = vt \cos(\theta - \alpha) = \frac{2v^2 \cos(\theta - \alpha)}{g \cos \alpha} \sin \theta$$

$$= \frac{v^2}{g \cos \alpha} [\sin(2\theta - \alpha) + \sin \alpha]$$

since $2 \sin A \cos B = \sin(A + B) + \sin(A - B)$. The range on the inclined plane is given by $x/\cos \alpha$, so that

$$R = \frac{v^2}{g \cos^2 \alpha} [\sin(2\theta - \alpha) + \sin \alpha].$$

The maximum values of R and of x, R_m and x_m, obviously occur when $\sin(2\theta - \alpha)$ has its maximum value for $(2\theta - \alpha) = \tfrac{1}{2}\pi$ when $\sin(2\theta - \alpha) = 1$, so that

$$R_m = \frac{v^2}{g \cos^2\alpha}[1 + \sin \alpha] \qquad \text{and} \qquad x_m = \frac{v^2}{g \cos^2\alpha}[1 + \sin \alpha].$$

We may evaluate the value of α corresponding to R_m since, from Fig. 7.4, $\sin \alpha = h/R_m$ and so

$$\sin \alpha = \frac{hg \cos^2 \alpha}{v^2(1 + \sin \alpha)}.$$

On multiplying out the denominator and transposing the terms we obtain an equation quadratic in $\sin \alpha$,

$$\sin^2\alpha + \frac{v^2}{(v^2 + hg)} \sin \alpha - \frac{gh}{(v^2 + hg)} = 0,$$

the solution of which is

$$\sin \alpha = gh/(v^2 + gh).$$

By using the fact that $\cos^2\alpha = 1 - \sin^2\alpha$, we obtain

$$\cos \alpha = \frac{v}{(v^2 + gh)} \sqrt{(v^2 + 2gh)},$$

leading to

$$x_m = (v/g)(v^2 + 2gh)^{1/2} \quad \text{and} \quad R_m = (v^2 + 2gh)/g.$$

To proceed further, recollect that the relation between the angle of inclination, α, of the plane and the optimum angle of projection, θ_m, relative to the plane, $2\theta_m = (\frac{1}{2}\pi + \alpha)$. This equality means that $\cos 2\theta_m = -\sin \alpha$ and, since $\cos 2\theta = 1 - 2\sin^2\theta$,

$$\sin^2\theta_m = \tfrac{1}{2}[1 + \sin \alpha]$$
$$= \tfrac{1}{2}[1 + \{gh/(v^2 + gh)\}],$$

so that

$$\sin \theta_m = \tfrac{1}{2}\left(\frac{v^2 - 2gh}{v^2 + gh} \right)^{1/2}$$

and finally, the angle of projection relative to the horizontal, $\phi_m = (\theta_m - \alpha)$, is given by

$$\sin \phi_m = \sin(\theta_m - \alpha) = \frac{v}{\sqrt{2(v^2 + gh)}}.$$

An obvious corollary to this result is that, since $2\theta_m = \frac{1}{2}\pi + \alpha$, the optimum angle of projection bisects the larger angle between the plane and the vertical.

If the projectile is fired up the plane an exactly similar procedure, but with $h < 0$, leads to equations which are similar to those deduced above, but with some plus signs replaced by minus signs. As an example, when $h < 0$, $2\theta_m = \frac{1}{2}\pi - \alpha$. This time, the optimum angle of projection bisects the smaller angle between the plane and the vertical.

These equations are quite consistent with those derived for projection above a horizontal plane for, if $\alpha = 0$, $R = (v^2/g)\sin 2\theta$, $\sin \phi_m = \sin \theta_m = 0.707$, etc. The important difference between the cases of the horizontal and inclined planes is that the optimum angle of projection is 45° only for the horizontal plane. Otherwise $\phi_m < 45°$ for projection down the plane or $\phi_m > 45°$ for projection up the plane.

Exercises

7.1. Two springs AB and CD, resting on a smooth horizontal surface, have their ends A and D attached to opposite points on two rigid walls separated by 0.8 m. The springs have unstretched lengths $l_{AB} = 20$ cm and $l_{CD} = 30$ cm, and spring constants $k_{AB} = 4$ kN m^{-1} and $k_{CD} = 2$ kN m^{-1}. The ends B and C of the springs are joined together. What will be the position of the junction BC when the system attains equilibrium? A mass of 2 kg is attached to BC and displaced by 5 cm. What would be the initial acceleration of the mass if it were released from this position?

7.2. Two masses of 1 kg are attached to the ends of a light string which passes over a massless pulley mounted on a horizontal bearing. If the string is not long enough for either mass to be in contact with the ground and its central point is vertically above the pulley bearing, what is the tension in the string? What difference would it make if the central point of the string were level with the pulley bearing? A mass of 0.2 kg is added to one of the 1 kg masses. What then is the tension in the string?

7.3. Inside an evacuated enclosure, a charged particle (which has a mass m and carries a charge e) is travelling with a velocity v_x along a straight line, the x-axis, when it enters the uniform electric field, directed along the y-axis, produced by two parallel plates the potential difference of which is V volts and the separation between which is d m. What force acts on the particle when it enters the field? What path does the particle follow while it is in the region of the uniform electric field?

7.4. A toy vehicle of mass 1 kg is pushed across a smooth horizontal floor. It contains a motor which can deliver a constant power of 10 W. If the motor is started when the velocity of the vehicle is 1 ms^{-1} (in the direction in which the motor will drive it), what is the force acting on the vehicle at that time? What will be the time interval before the vehicle has a velocity of 2 ms^{-1}?

7.5. The spring buffers on a model railway are compressed by 2.5 cm when a force of 40 N is applied to them. If a model railway engine of mass 0.25 kg freewheels into the buffers with a velocity of 0.9 ms^{-1}, how far will the buffers be compressed before the engine comes to rest?

7.6. An electron of charge 1.6×10^{-19} C and mass 9×10^{-31} kg is emitted at the cathode of a television tube. If the potential difference through which it accelerates towards the screen is 7 kV, what is its velocity when it reaches the screen?

7.7. A proton, the specific charge e/m of which is 10^8 C kg^{-1} is accelerated through a potential difference of 2 V. What is its velocity? After its acceleration the proton passes through a hole in a plane plate A into a uniform electric field of $10\,V\,m^{-1}$ produced by a second plate B which is parallel to plate A and separated from it by 0.25 m, having a potential different from that of A. If the proton's velocity is directed at right angles to the equipo-tential lines, is it possible for the proton to impinge on the plate B? If the acute angle between the direction of the proton's initial velocity (when it enters the field) and the equipotential lines is α, and the proton cannot reach plate B, what is the maximum distance it will travel parallel to the plates before it collides with plate A? If this range is required to be 12 cm, at what value of the angle α would the proton enter the field?

7.8. A railway carriage of mass 10 tonnes, moving with a speed of $1.2\,m\,s^{-1}$, collides with a similar carriage which is stationary but free to move, and couples with it. The buffer springs of the carriages, obeying Hooke's Law, have an effective spring constant so that a force of 0.5 MN between the carriages decreases their separation by 25 cm. What is the greatest compres-sion produced in the springs?

7.9. The base line of a tennis court is 11.5 m from the net which is 90 cm high. The server strikes the ball so that its initial velocity is directed horizontally when it is 2.4 m above the ground. What is the minimum velocity with which the ball must be served so that it clears the net? For a service to be good the ball has to strike the ground, at the most, at 17.5 m from the base line. What is the maximum value of horizontally directed initial velocity of the ball that will keep the line judges silent? Will this high speed service clear the net?

7.10. A shot putter impels the shot, which has a mass of 7.25 kg, with a velocity of $13.5\,m\,s^{-1}$ from a height of 2 m above the ground (which is horizontal). What is the direction of projection of the shot, relative to the ground, which would maximize its range? What is the maximum range in this case?

Solutions

7.1. The walls are separated by 80 cm and the relaxed length of the two springs is $(20 + 30) = 50$ cm. The sum of the extensions, x_A and x_C, of the two springs must be 30 cm; that is, $x_A + x_C = 0.3$. When equilibrium has been attained the forces exerted by the springs at their junction are equal and oppositely directed so that $(k_A x_A - k_C x_C) = 0 = (4 \times 10^3 x_A - 2 \times 10^3 x_C)$, that is, $x_C = 2x_A$, with the result that $x_C = 0.2$ m and $x_A = 0.1$ m. The equilibrium position of the junction BC is $(0.2 + 0.1) = 0.3$ m from A.

Say that the junction BC (with the 2 kg mass attached) is displaced from its equilibrium position by a distance y towards A. The tension in AB is then reduced to $F_A = k_A(x_A - y)$, while that in CD increases to $F_C = k_C(x_C + y)$. The difference between these is the force acting on the 2 kg mass:

$$F = (k_A x_A - k_C x_C) - (k_A + k_C)y = -(k_A + k_C)y.$$

This leads to the initial acceleration $a = \frac{1}{2}(6 \times 10^3 \times 0.05) = 150\,m\,s^{-2}$.

It's as well to notice here that if we visualize the springs AB and CD as a combination (as in Example 25) the spring constant of the combination appears to be $K = (k_A k_C)/)(k_A + k_C)$; that is, we have a series combination in which $F = F_A = F_C$ and $F = 0.3 \times K$. The way in which the spring constants combine in the derivation of the equation of motion of the 2 kg mass at the junction BC, $K = k_A + k_C$, is that of the parallel combination of springs in Example 25. This apparent paradox comes about because in the non-equilibrium state (when the mass is moved) the change in length is the same for both springs. The fixed points A and D could be imagined as superposed, with the springs in parallel. When the initial equilibrium state is created the process could be imagined as fixing A to its wall, joining the springs at BC, and stretching the combination to reach the other wall. Obviously, in this case the springs, the individual extensions of which are different, then behave as if they formed a parallel combination.

7.2. Take as an illustration a single mass, m, supported from a fixed point by a light string hanging vertically. There is a force acting downwards on the system, the weight of the mass. Since the string supports the mass in an equilibrium state this downward force must be nullified by an upward tension in the string, the magnitude of which is that of the weight, tension $T = mg$. This tension produces a reactive force mg at the point of support of the system, so that the string serves to make the point of support the point where a force mg acts.

When the string passes over a pulley the downwards forces are the weights of the masses. The string will hang vertically from the edges of the pulley, at its left and at its right with the centre of the string, say, at the highest point of the pulley. If the tension in the string is T, then at the left the magnitude of the force acting on the string–mass system is $mg - T$ directed away from the string centre. At the right there is also a force of magnitude $mg - T$ directed away from the string centre. These two forces have equal magnitudes and opposite directions (although they both act downwards) so that there is zero force overall acting on the system. It is in equilibrium and remains in its state of rest (or of uniform velocity). In symbols we write $mg - T = -(mg - T)$, or $T = mg$. Obviously, when the two masses are equal the state of equilibrium is maintained. The numerical value of T for $m = 1$ kg is 9.8 N.

The addition of an extra mass to one end of the string means that there will be a net force acting on the system. Say that the 0.2 kg mass is added at the left end of the string so that it carries a mass $m_1 = 1.2$ kg. The extra mass will upset the equilibrium $[m_1 g - T \neq -(mg - T)]$ and there will be a net force acting. On the left-hand side of the pulley this leads to the equation of motion

$$m_1 g - T = m_1 a$$

where a is the acceleration away from the equilibrium state, while on the right-hand side

$$mg - T = ma,$$

so that

$$\frac{T}{m} + \frac{T}{m_1} = 2g \quad \text{or} \quad T = \frac{2m_1 m_2}{m + m_1} g$$

which, for $m = 1$ kg, leads to $T = 1.09g = 10.7$ N, and to $a = 0.9 \, \text{m s}^{-2}$.

7.3. The electric field between the two parallel plates is $E_y = -dV_E/dy$ where y is an axis normal to one of the plates. In this case the field can be taken as uniform and, if d is the separation of the plates, $E_y = V_E/d$ or $-V_E/d$ depending on the choice of direction for $y > 0$. The force acting on the charge is $E_y q$ along the y-direction so that the acceleration in the y-direction is $a_y = E_y q/m$. If we take the initial point on the straight line path of the particle to have the y-coordinate $y = 0$, then at time t after the particle has entered the field $y = \frac{1}{2}at^2$. The initial velocity component perpendicular to y, v_x, remains unchanged because $E_x = -(dV_E/dx) = 0$, with the consequence that $a_x = 0$. Choosing the edge of the electric field in the x-direction as $x = 0$, we may say that $t = (x/v_x)$ and

$$y = \tfrac{1}{2}a_y(x/v_x)^2 = \frac{1}{2}\frac{q}{m}\frac{E_y}{v_x^2}x^2$$

showing that the path of the particle in the field is a parabola. When the particle leaves the field the electrical force will no longer act on it and it will continue to travel along a straight line which has the same slope as the parabola at the point at which the particle left the field. An equation derived from the one obtained above was used in the analysis of J. J. Thomson's experiment to measure the specific charge e/m of an electron.

7.4. Let's say that time is measured from the instant when the motor is started, so that $t = 0$ when $v = 1$. The power is the rate of working, $P = dW/dt$, so that for a constant power the force acting when the velocity is v is $F = P/v$. For $P = 10$, $v = 1$, and $F = 10\,\text{N}$. In this case the rate of working is also the rate of change of kinetic energy which changes by $\delta T = \delta W = \frac{1}{2}[1 \times (4 - 1)] = 3/2\,\text{J}$ as the velocity increases to $2\,\text{m s}^{-1}$:

$$\delta W = \frac{dW}{dt}\,\delta t = 10\delta t, \qquad \text{giving} \qquad \delta t = 0.15\,\text{s}.$$

7.5. The spring constant of the buffers is obtained from the 'static deflection' information. $F = kx$, or $k = 40/0.025 = 1600\,\text{N m}^{-1}$. The model engine will come to rest when its kinetic energy has been converted all to potential energy, giving $\frac{1}{2}mv^2 = \frac{1}{2}kx^2$, or $x^2 = mv^2/k$, leading to $x = 1.125\,\text{cm}$.

7.6. The potential difference of $7\,\text{kV}$ corresponds to a potential energy difference (for a charge q coulombs) of $7q\,\text{kJ}$. Inside the evacuated television tube, in the absence of collisions with gas molecules, the potential energy is converted entirely to kinetic energy, $qV_E = \frac{1}{2}mv^2$, or $v^2 = 2qV_E/m$, giving $v = 1.9 \times 10^7\,\text{m s}^{-1}$. Alternatively, we might have expressed our result by saying that the velocity, v, attained by an electron accelerated through a potential difference V_E, is given by $v = 5.93 \times 10^5 V_E^{1/2}\,\text{m s}^{-1}$.

7.7. The change in potential energy of the proton is equal to the increase of its potential energy when it passes through a potential difference of $2\,\text{V}$. If its charge is Q, the potential difference through which it travels is V_E, and its mass is m, then $V_E Q = \frac{1}{2}mv^2$, or $v = (2V_E Q/m)^{1/2}$: (Q/m) is the specific charge of the proton, and so

$$v = (4 \times 10^8)^{1/2} = 2 \times 10^4\,\text{m s}^{-1}$$

The potential difference, V_E, between the plates, which are separated by a distance d, producing the uniform electric field E_y is $V_E = E_y d = 2.5\,\text{V}$. If this potential is an accelerating potential (the potential, V_{EC}, of the plate C is greater than V_{EA}, that of plate A) the proton is bound to reach the plate C because the field between the plates will have the same sign as the field which accelerated the proton. It will arrive then at plate A with a velocity $v = (9 \times 10^8)^{1/2} = 3 \times 10^4\,\text{m s}^{-1}$. On the other hand, if $V_{EC} < V_{EA}$ the potential difference will have a retarding effect and, since $2.5 > 2$, there is no possibility that the proton can impinge on plate A. The distance of closest approach of the proton to plate C will be 5 cm.

If the angle between the direction of the initial velocity of the proton and the equipotential lines is α then the parametric equation for the displacement parallel to the lines of electric field will be

$$y = vt \sin \alpha - \frac{E_y Q}{2m} t^2,$$

which is zero for

$$t = \frac{2m}{E_y Q} v \sin \alpha = 2 \times 10^{-9}(2 \times 10^4 \sin \alpha) = 4 \times 10^{-5} \sin \alpha.$$

The distance travelled parallel to the equipotential lines in this time will be $R = vt \cos \alpha = 2 \times 10^4(4 \times 10^{-5} \sin \alpha \cos \alpha) = 0.4 \sin 2\alpha$. The maximum value of R will be 0.04 m. The path of the proton is parabolic, of course. If $R = 0.12 = 0.4 \sin 2\alpha$ then $\sin 2\alpha = 0.3$ and $\alpha = 8.73°$ or $81.27°$, since $\sin 2\alpha = \sin(\pi - 2\alpha)$. The two angles of projection, α and α', required to produce a given range for a particular velocity of projection, v, have to be complementary angles $\alpha + \alpha' = 90°$.

7.8. In the collision, momentum will be conserved. The final velocity is derived from

$$10^4 \times 1.2 = 2 \times 10^4 v, \qquad \text{that is} \quad v = 0.6\,\text{m s}^{-1}.$$

The difference in kinetic energy before and after coupling of the carriages is

$$\tfrac{1}{2} \times 10^4 \left[(1.2)^2 - \tfrac{1}{2}(2 \times 10^4) \times (0.6)^2\right] = 0.36 \times 10^4\,\text{J}.$$

Making the assumption that this kinetic energy has been transformed entirely to potential energy in the springs, we may say that

$$0.36 \times 10^4 = \tfrac{1}{2}kx^2 = \tfrac{1}{2}[(5 \times 10^5)/0.25]x^2,$$

leading to $x = 19\,\text{cm}$.

In reality, some of the kinetic energy will be transformed to sound energy, even if we are able to neglect the effect of forces resisting the motion of the trucks. The method used above is a first approximation only. It is justifiable because the quantity of kinetic energy transformed to sound energy is relatively small, probably of the order of one joule or less, even when the sound is quite intense.

Neglect of the resistive forces is likely to have a much greater effect on our result than is ignoring the production of noise.

7.9. To clear the net the ball must fall through a distance less than 1.5 m in the time interval, t, during which the ball reaches the net. The maximum value that will be allowed to t is given by $t^2 = 2(1.5/g)$, or $t = 0.553$ s. In this time interval the ball travels through a horizontal distance of 11.5 m, so that its initial velocity must be $v = 11.5/0.553 = 20.8\,\mathrm{m\,s^{-1}}$.

Will this service be good? This will only be the case if the vertical distance travelled in the time, t_2, taken to cover 17.5 m horizontally is more than 2.4 m. The time to travel 17.5 m is $t_2 = 17.5/20.8 = 0.841$ s. The vertical distance travelled in this time is $Y = \frac{1}{2}gt_2^2 = 3.47$ m. The service is 'in'.

For the case of maximum velocity we require in the time interval t_2 that $Y = 2.4$ m; that is, that $t_2^2 = (2 \times 2.4/g)$, or $t_2 = 0.75$ s. This means that $v_m = 17.5/t_2 = 25\,\mathrm{m\,s^{-1}}$. In this case the ball will clear the net by 46 cm.

This result is perhaps surprising, because we might expect the maximum velocity of projection of a tennis ball in a service to be nearer to $50\,\mathrm{m\,s^{-1}}$ than to $20\,\mathrm{m\,s^{-1}}$. The difference between these two values can be explained in terms of the direction of the initial velocity. Say that our server increases the initial velocity of the ball, this time jumping off the ground so that the initial height of the ball is 2.7 m. Let the direction of the initial velocity be along a line inclined 7° below the horizontal. The displacement of the ball relative to $t = 0$ is then given by

$$x = 50t \cos 7°, \qquad -y = -50t \sin 7° - \tfrac{1}{2}gt^2.$$

The ball reaches the net when $x = 11.5$ m, defining t, and at that time $y = 1.68$ m. The ball passes 12 cm above the net. For $x = 17.5$, the value of y is 2.76 m, so that the service is good. The downward component of the initial velocity adds to the velocity gained by falling under the influence of gravity, so that the ball reaches the ground sooner than it would in free fall.

In practice, the trajectory of the tennis ball is unlikely to be nearly parabolic because of the resistive forces due to its passage through the atmosphere, and because it is likely to have spin imparted to it in the action of serving.

7.10. This exercise can be imagined to be one of projection of a projectile down a plane, the inclination of which to the horizontal is $\alpha = \tan^{-1}(h/x)$, where h is the height of projection and x is the horizontal range. Then, using the formulas derived for projection down an inclined plane, the angle of projection, ϕ_m, for maximum range is given by $\sin \phi_m = \sin(\theta_m - \alpha) = v[2(v^2 + gh)]^{-1/2}$. With $v = 13.5$ and $h = 2$, $\phi_m = 42.41°$ and $x_m = [(v/g)(v^2 + 2gh)^{1/2}] = 20.48$ m. The calculation could have been reduced a little by calculating first x_m and then finding α. Then

$$2\phi_m = (\pi/2) - \alpha.$$

It is interesting to notice that a 12% reduction in the height of projection (again with $v = 13.5\,\mathrm{m\,s^{-1}}$) decreases x_m only by about 1%, while a change in velocity of only 4% alters the maximum range by about 7%. Small changes in the angle of projection, say of about 5%, make a difference only of about 1% in x for given values of h and v. The putter should pay most attention to imparting the highest possible velocity to the shot in a direction close to the optimum direction. The most important criterion is that the impulse given to the shot by the athlete should be the largest possible.

We have neglected dissipative effects in the example of the shot putter with more justification than when we dealt with the tennis service in Exercise 7.9. This is excused by the observations that the density of the shot is large and that the velocity of the shot is relatively low.

8
Energy dissipation

8.1 Resistive forces

So far, we have encountered dissipative forces which provide resistance to motion, giving, for example, an upper limit on the velocity of a ship. In the case of a body moving in a fluid it isn't difficult to visualize the origin of such forces when we recollect that a force is exerted on a wall by a fluid jet, say with cross-section A and velocity v, incident upon it, as shown in Fig. 5.5. The mass of the fluid reaching the wall in unit time is $Av\rho$ where ρ is the fluid density, and this mass of fluid has a momentum $p = \rho A v^2$. If the fluid jet loses all its momentum in the 'collision' with the solid wall, the rate of change of momentum is simply $\rho a v^2$; in other words, the force acting on the wall, $F \propto v^2$. Alternatively, we could say that the wall exerts a force F on the fluid jet to decelerate it and the force acting on the wall is the reaction to this force at the wall. If the fluid were static and the wall (as a body) moved through it with velocity v, we could make a similar qualitative argument to justify the observation that the force resisting the motion of the wall would be proportional to the square of its velocity, $F \propto v^2$. This force comes about because of the exchange of momentum between the body (the wall) moving through the fluid and the fluid itself. In the case of the fluid jet incident on the wall the fluid, if it lost all its momentum, would be incident on the wall and then simply slip down its surface. In fact, the impact of a fluid jet on a wall is much more complicated than that; some of the fluid slips down the surface, some is broken into droplets which appear to rebound from the wall, and some may be reflected more or less as a jet. Our simplified treatment of this phenomenon provided us with an estimate of the force exerted on the wall by the fluid incident on it, and took no account of the way in which the cylindrical water jet breaks up as it strikes the wall. In the same way, we have ignored the complexities of the fluid behaviour which will accompany the movement of the body through the fluid. In both cases we make an assumption that the force is proportional to the square of the velocity as a first reasonable approximation to a problem which is very complicated in detail.

This qualitative argument justifies our assuming that when a body moves through a fluid with such a velocity that it causes considerable agitation of the fluid— as, for example, a ship leaving behind it an extensive wake when sailing on a calm sea —the resistive force is likely to be related to the square of the body's velocity. What if the body moves very slowly through the fluid so that, in effect, it slips through the water and causes no agitation of the fluid? In this case the body, as it moves along, first pushes the fluid aside and then pulls it back into place, so that the fluid behind the body has the same velocity relative to the body as does the fluid in front of it. The momentum gained by the fluid as it is pushed aside is compensated by the momentum lost by it when it moves back into place. The resistive force is not calculable in terms of the fluid momentum in front of and behind the body. It can be shown that the resistive force acting on a sphere falling slowly through a fluid is proportional to its velocity. It seems not unreasonable to extend this result so that we assume, in the case of low speed motion of a body in a fluid, that the force of resistance to the motion of the body will be proportional to the velocity, v, of the body, $F \propto v$.

The regimes of v for which $F \propto v$ and $F \propto v^2$ depend on the properties of the fluid and on the shape of the body. When the fluid is dense and incompressible (a liquid) the linear regime is restricted to small velocities even for a smoothly surfaced, well-shaped body. When the body slips through the fluid in this low velocity regime, the behaviour of the fluid is described as streamline flow.

When the body moving through the fluid experiences a force of resistance to its motion, it is being subjected to a drag force, which is an inevitable corollary to movement through a fluid. The drag force depends on the body, the fluid, and their relative velocity. The two very simple possibilities of the relationship between drag force and velocity outlined above are often quite useful as first approximations to real problems. The choice of the linear or parabolic velocity regime is usually made empirically. Because the force is velocity dependent, the equations relating velocity with displacement or with time are no longer as simple as in the case in which the force is constant.

8.1.1 Terminal velocity

One interesting point about drag forces related to the velocity is that when they act in opposition to a constant force, F, which is accelerating a body of mass m, the net force acting on the body when its velocity is v is $F_n = F - mkf(v)$, where k is a constant and $f(v)$ is the velocity dependence of the drag force, which may be linear or parabolic, or even more complicated. As the velocity increases, so F_n becomes smaller, until $F_n = 0$ when $f(v) = (F/mk)$. The body then has no force acting on it and continues to travel with the limiting velocity which is the velocity at which F_n becomes zero. This limiting velocity is also known as the terminal velocity.

8.2 Friction

The idea of forces resisting motion is not restricted only to the case of a body travelling through a fluid. When we attempt to pull or push moderately heavy articles across floors which are not smooth, we find that we cannot do so except by the expenditure of considerable effort. The forces against which we have to push are forces due to friction between the floor and the article we are pushing. In the case in which we are pushing or pulling bodies we can say that frictional forces are the forces resisting motion (parallel to a solid surface) of solid objects and have their origin in the contact made between the solid objects and the surface on which they move or stand still. If we are to succeed in moving the object, then the soles of our shoes in contact with the floor must not slip; otherwise, we will be moving only our limbs rather than pushing with any effect on the body. If the frictional force due to contact between shoes and floor is greater than that between the object and the floor, it is possible that the object may be moved, provided that the effort required does not exceed our strength. In terms of the pusher, the frictional force is the resistance to his motion which gives him the ability to push with forces less than the frictional force between his shoes and the floor. Alternatively, we may say that the force which drives a car forward is the frictional force between its tyres and the ground. Power is supplied to the wheels by the engine and transmission, and if it is supplied so that the force exerted by the tyres exceeds the frictional force the tyres slip (in just the same way as the soles of our shoes would slip). The vehicle spins its wheels and the engine power is not used fully.

The phenomena of frictional forces are described in terms of the force which the surface exerts in reaction to the weight of the solid object, and two coefficients of friction, the static coefficient of friction, μ_s, and the kinetic coefficient of friction, μ_k. If we push a solid object resting on a plane floor with increasing force we will find that it remains still until the force component (with which we push) parallel to the surface reaches a certain critical value, as illustrated in Fig. 8.1. The reaction of

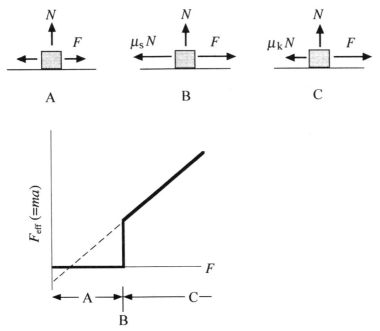

Fig. 8.1 A graph showing the 'effective force', F_{eff} ($= ma$, where m is the mass of the body and a is its acceleration), acting on a body which produces a normal force N when it rests on a horizontal plane, as a function of the horizontally directed force, F, applied to it. The coefficients of static friction and of sliding friction between the body and the plane are μ_s and μ_k, respectively. The graph is in two parts, the range A for which $F < \mu_s N$ so that the body does not move, and the range C for which $F > \mu_s N$. In range C the equation of motion takes the form $ma = F - \mu_k N$. In range A the force resisting the motion is always equal to the applied force; otherwise the body would move in the direction opposed to F. Force diagrams for the ranges A and C are given above the graph, and the transitional case, B, for which $F = \mu_s N$ is shown for completeness.

the surface to the weight of the object is the normal force, N (normal to the surface), and the force, F_s, required to initiate movement defines the coefficient of static friction, μ_s, by $F_s = \mu_s N$, with $F = |F|$ and $N = |N|$. Once the object has started to move it is found that the force, F_k, required to keep it moving with constant velocity is less than or equal to F_s, and is pretty well constant: F_k is used to define the coefficient of kinetic friction, μ_k, by $F_k = \mu_k N$. Because $F_k \leqslant F_s$ while N is the same, whether or not the object moves, it can be said that $\mu_k \leqslant \mu_s$. The forces F_s and F_k refer to a particular object standing or sliding on a particular surface under a particular set of conditions. Obviously, the cleanliness of the surfaces, their state of polish or smoothness, the temperature, and the relative velocity of the object and surface could all have a marked effect on the magnitudes of the coefficients of friction. Both μ_s and μ_k can be greater than 1, but it's usual for them to be less than 1.

The force opposing the motion along the surface of the plane is directed parallel to the surface, while the normal force is directed along a normal to the surface. The two equations defining μ_s and μ_k are scalar equations, because the μ's are constants and the resistive forces are directed perpendicular to the N's. The resistive force produced by kinetic friction is constant only as long as μ_k is constant, which might not be appropriate if the temperature of the surface–object interface increased markedly. Since the frictional forces are dissipative and transform mechanical work into forms of energy other than kinetic or potential, this last situation seems quite likely to happen.

Here we have considered only two classes of mechanical dissipative forces, constant or velocity dependent. In the case of a solid moving on a solid surface the force resisting its motion is near enough constant provided that conditions don't vary too much. Given a particular driving power, the body will accelerate, but the driving force derived from the driving power will be reduced as the body's velocity increases. The body has its velocity increased until the frictional force resisting its motion is equal to the driving force. Then, in the absence of any other forces, the body will be in a state in which its velocity is constant. It will have attained its maximum, limiting, or terminal, velocity. If the driving force were constant there would be no limit to the velocity of the body, provided that the frictional force did not exceed the driving force (in which case the body would not move). When the dissipative force is velocity dependent the velocity increases under the influence of an accelerating force until the dissipative force becomes as large as the driving force, when a state of uniform velocity is assumed. The limiting velocities have different origins in that for the constant frictional force the driving force (for a constant power output) decreases to become equal to the resistive force, while for velocity dependent dissipative forces the dissipative force increases to become equal to the driving force.

8.2.1 Energy conversion

How are these forces dissipative? To provide an answer we need to look at the origins of the forces. Friction between solid surfaces occurs because the surfaces are rough, and may be reduced by polishing one or both surfaces smooth. In a qualitative way we could explain the difference in effort required between striking an ordinary match on sandpaper and striking a safety match on its striking surface as due in the first instance to the difference between the roughness of the sandpaper and the relatively smooth striking surface used with the safety match. Heat is generated when the match is forced across the sandpaper and this leads to ignition. When the sandpaper is well used, smoothed by the debris trapped between its grains, it becomes much more difficult to ignite the match. The dissipation due to friction is due to the conversion of mechanical energy to heat energy and sound energy. In a similar way, with velocity dependent forces (often called viscous forces) the dissipation is due to the transformation of mechanical energy by the transfer of kinetic energy of the moving body to the fluid in which it moves and the subsequent conversion of this transferred kinetic energy to heat and sound energy.

Example 42.   A cube of wood of mass 10 kg lies on a wooden table which forms a horizontal plane. If the coefficient of static friction of wood on wood is 0.27, what is the magnitude of the horizontal force applied to the wood block which will cause it to move? If the coefficient of kinetic friction is 0.25, what will be the acceleration of the block once it starts to move, assuming the horizontal force remains constant?

The normal force is simply $mg = 98.1$ N. The force required to initiate motion will be $F = 0.27 \times 98.1 = 26.49$ N. Once the block is moving the resistive force decreases to $F_1 = 0.25 \times 98.1 = 24.53$ N. The force producing acceleration will be 2.04 N. Rounded up to the first decimal place, the acceleration of the block will be $0.2 \, \text{m s}^{-2}$.

It's of interest to notice that the motion of the block could be produced by lifting one end of the table to produce a sloping rather than a horizontal surface, even in the absence of an external force acting parallel to the plane. Then, if the angle of tilt is θ, the force down the slope acting on the cube will be $mg \sin \theta$, the normal force will be $mg \cos \theta$, and the frictional force acting up the slope will be $\mu_s \, mg \cos \theta$. When these forces are equal in magnitude, $\tan \theta = \mu_s$, which is independent of the mass of the block. Once the wood block begins to slip the force acting on it down the slope will be $(mg \sin \theta - \mu_s \, mg \cos \theta)$, so that the acceleration of the block will be $g(\sin \theta - \mu_s \cos \theta)$, and the block will continue to accelerate for as long as the slope continues. On the other hand, if $\mu_k < \mu_s$, the slope of the surface could be reduced until the block moved with constant velocity down it. In such a case, if the angle for constant velocity of the block were ϕ, we would have $\tan \phi = \mu_k$.

Example 43.   A motor car of mass 1 tonne is travelling at $25 \, \text{m s}^{-1}$ along a level road. If the coefficient of friction between a tyre and the road surface is 0.6, what is the minimum distance in which the car can be brought to a halt?

For the stopping distance to be a minimum the driving force of the car's engine must be zero; that is, the clutch is depressed. The only force acting in the direction of motion is the frictional force $F = -\mu N = -\mu mg$, or $F = -5.88$ kN. The retardation produced is $a = F/m = -5.88 \, \text{m s}^{-2}$, leading to the minimum value of the stopping distance $x = -v^2/2a = 53.2\,\text{m}$. The stopping distance is independent of the mass of the car.

Obviously, the coefficient of friction plays an important role in determining the stopping distance. If the road had a wet surface so that the coefficient of friction were reduced to 0.3, the stopping distance would be doubled. If the car were sliding with its wheels locked, the coefficient of kinetic friction would determine the stopping distance, while if they continued to rotate the coefficient of static friction would be the relevant one. Since usually $\mu_s > \mu_k$, the advantage of anti-lock braking is obvious.

Example 44. A rope will break if it is subjected to a tensile force of more than 1.5 kN. What is the greatest mass of a load that can be dragged by the rope across a horizontal floor if the coefficient of friction between the load and the floor is 0.3 and the rope is horizontal? Can this greatest load be increased without altering the rope?

When the rope is horizontal the normal force is the weight of the load, mg, and movement of the load requires a horizontal force greater than the frictional force μmg, when μ is the coefficient of friction. This force must not be greater than the breaking tension of the rope. An estimate of the greatest load can be obtained by equating the breaking tension to the frictional force; that is,

$$1.5 \times 10^3 = 0.3 \times 9.8m, \qquad \text{giving} \qquad m = 510\,\text{kg}.$$

This maximum load can be increased without altering the rope. Referring to Fig. 8.2, it's seen that when the rope is inclined at an angle θ to the horizontal, the normal force is reduced by $T\sin\theta$, where T is the tension in the rope. This reduces the frictional force so that the equation for the greatest load takes the form

$$\mu(mg - T\sin\theta) = T\cos\theta;$$

that is,

$$m = \frac{T}{\mu g}(\cos\theta + \mu\sin\theta).$$

The load will exceed 510 kg provided that

$$(\cos\theta + \mu\sin\theta) > 1.$$

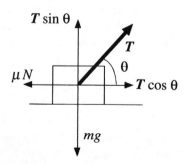

Fig. 8.2 The forces acting on a load pulled by a rope inclined at an angle to the horizontal. The load has mass m and the tension in the rope is $\mathbf{T}$. The coefficient of friction is μ.

It's easy to verify that this condition is satisfied for $\theta < 33.4°$.

The upper limit of load which can be moved by the rope is found from the maximum value of $(\cos \theta + \sin \theta)$. At the maximum, the slope of m plotted as a function of θ is zero, so that

$$\frac{dm}{d\theta} = \frac{T}{\mu g}(\mu \cos \theta - \sin \theta) = 0.$$

This condition is satisfied only for $\tan \theta = \mu$, and in this case $\theta = 16.7°$. The limiting load is then 532.5 kg.

Example 45. In the sport of curling, a stone of mass 7 kg is projected along an ice surface with the aim that it should travel to a mark some 20 m from the point of projection. If the coefficient of friction between the ice and the stone is 0.1, with what velocity should the stone be projected so that it reaches the mark without overshooting it?

In this case we may use energy conservation to find the velocity. The kinetic energy of the stone is initially $\frac{1}{2}mv^2$, and this energy is 'used up' in doing work against the frictional force. The frictional force is constant and so $\frac{1}{2}mv^2 = \mu mg \times x$, where x is the distance that the stone will travel. Substitution into this equation gives $v = 6.26 \, \text{m s}^{-1}$.

The kinetic energy of the stone has been transformed to other forms of energy; not, as we stated loosely, 'used up'. The total energy of the stone/ice system remains constant, although the individual energies of its components, the stone and the ice surface, alter considerably.

Example 46. A falling raindrop has its movement through the air resisted by a force which is proportional to its velocity. If the maximum velocity of the falling raindrop is $9.8 \, \text{m s}^{-1}$, at how many seconds into its fall will the raindrop attain this velocity?

The resistive force is proportional to v, so that we may write it as $F_r = -mkv$, where m is the mass of the raindrop and k is a constant (we could have said that the resistive force was proportional to the momentum). The accelerating force acting on the raindrop is its weight, so that the total force is

$$F = F_a - F_r = m(g - kv).$$

The downwards acceleration of the raindrop is then simply $F/m = g - kv$. The terminal velocity of the raindrop occurs when the force acting on the raindrop is zero; that is, for $v = V_T = g/k$. The acceleration of the drop may be written as

$$dv/dt = k[(g/k) - v] = k[V_T - v].$$

The integral form of this equation is

$$\int \frac{dv}{V_T - v} = k \int dt$$

or

$$-\ln(V_T - v) = kt + C,$$

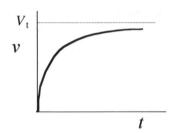

Fig. 8.3 Velocity, v, as a function of time, t, for a raindrop subject to a resistive force linear in v while falling under the influence of gravity.

where C is the integration constant. Since $v = 0$ when $t = 0$, $C = -\ln(V_T)$, and

$$-kt = \ln[(V_T - v)/V_T]$$

or

$$v = V_T[1 - \exp\{-kt\}] = V_T[1 - \exp\{-(gt/V_T)\}].$$

This equation is illustrated graphically in Fig. 8.3. For $V_T = 9.8$,

$$v = 9.8[1 - \exp\{-t\}]$$

and, strictly speaking, the raindrop will attain its terminal velocity only when t is infinite. On the other hand, as we can see from Fig. 8.3, the velocity soon becomes very close to V_T. For the raindrop, when $t = 2\,\mathrm{s}$, v is only 14% less than V_T; by 4 s it is 2% less, by 6 s it is within 0.2% of V_T, and by $t = 10\,\mathrm{s}$ it is only 0.005% less than V_T. It seems fair to say that the raindrop reaches its limiting velocity after it has been falling for about 7 s. The time constant, $\tau = (V_T/g)$, is the time taken for the velocity to increase from zero to $0.63V_T$. The time required to attain a velocity within 0.1% of V_T is given by $t = 7\,\mathrm{s}$.

Since the terminal velocity of the raindrop is small, it is possible that equating the resistive force to kv may provide a reasonably good approximation to the actual resistive force.

Example 47. A cannonball is projected vertically upwards with an initial velocity of v_0. If the air resistance to its progress is proportional to the square of its velocity, v, what will be the maximum altitude that it will reach?

If we take $F_r = -mkv^2$, the acceleration equation has the form

$$a = v\frac{dv}{dy} = -g - kv^2,$$

where y is the height parameter. If the cannonball were falling it would have a terminal velocity defined by $V_T^2 = g/k$, so that we may write

$$v\frac{dv}{dy} = -(g/V_T^2)(V_T^2 + v^2)$$

or, by separating the variables and integrating,

$$\int \frac{v\,dv}{V_T^2 + v^2} = -\frac{g}{V_T^2}\int dy.$$

The integral involving v is simple, because $d(v^2) = 2v \, dv$. The top line in the integral is simply $\frac{1}{2}[d(V_T^2 + v^2)]$, so that if we set $(V_T^2 + v^2) = z$, the integrand has the form $\frac{1}{2}[dz/z]$, so that the integral has the form $\frac{1}{2} \ln z$; that is,

$$\tfrac{1}{2} \ln(v^2 + V_T^2) = -(g/V_T^2)y + C.$$

When $t = 0$, $v = v_0$, so that $C = \frac{1}{2} \ln(v_0^2 + V_T^2)$ and

$$\tfrac{1}{2} \ln\big[(v^2 + V_T^2)/(v_0^2 + V_T^2)\big] = -(g/V_T^2)y.$$

The maximum height is attained when $v = 0$, and, rearranging the terms, this maximum altitude is given by

$$y = (V_T^2/2g) \ln\big[1 + (v_0^2/V_T^2)\big].$$

Exercises

8.1. A skier of mass 80 kg sets off down a slope inclined at 15° to the horizontal. The ski run is straight down the slope and 1 km long. If the coefficient of friction between the skier's skis and the snow surface is 0.1, how fast will the skier be travelling at the end of his run?

8.2. The coefficient of friction between the driving wheels of a small racing car and the road surface on which it runs is 0.6. The mass of the car is 1 tonne, and the power available from its engine is 120 kW. If the driver runs the engine at full power when he starts the race, what distance will it have travelled before this full power is available to drive the car? What will its velocity then be?

8.3. A motor scooter and its rider have a combined mass of 150 kg. The rider finds that he can drive at a maximum speed of 80 kph down an incline of 1 in 20, but only at 48 kph up the same incline. If the resistance to the motion of the scooter is proportional to the square of its velocity, what is the maximum speed that the scooter can maintain on the level? What is the power of the scooter?

8.4. A railway train of mass 300 tonnes has a maximum speed of 96 kph on a level track. The resistance to its motion varies as the square of its velocity and the power produced by its engine is 750 kW. What it starts from rest along a level track how far will it travel before reaching a velocity of 48 kph?

8.5. A freefall parachutist is subject to a resistive force proportional to the square of his velocity. When he falls in a prone position, with body horizontal and arms and legs extended, his terminal velocity is $53 \, \mathrm{m\,s}^{-1}$. He leaves an aircraft at an altitude of 3000 m and assumes the prone position almost immediately. What will be his altitude when his speed is within 1% of his terminal velocity? If there were no resistance to his fall, how fast would he be travelling at this altitude?

Solutions

8.1. The force acting to accelerate the skier is a combination of the component of his weight parallel to the slope and the frictional force which resists his motion. The component of weight parallel to the surface is $mg \sin \theta$ and the normal force is $N = mg \cos \theta$, so that the skier's acceleration is given by

$$a = g \sin \theta - \mu g \cos \theta$$

which, for $\theta = 15°$, is $1.59 \, \mathrm{m s^{-2}}$. Assuming that the skier started with an initial velocity of zero, his velocity at the end of the run (of length x) is $v = (2ax)^{1/2}$, which for $x = 1 \, \mathrm{km}$ takes the value $v = 56.4 \, \mathrm{m s^{-1}}$.

This answer is nonsensical because the world record speed of travel downhill on skis is about $55 \, \mathrm{m s^{-1}}$, and that's down a slope of about 40°. This absurdity arises because we have neglected the effect of air resistance. Because the skier moves with a reasonably high velocity, and because he has an irregular shape, we may assume that the skier is subject to a resistive force, F_r, proportional to the square of his velocity; $F_r = -mkv^2$. Then

$$a = g(\sin \theta - \mu \cos \theta) - kv^2$$

which, for a fixed value of θ, may be written $a = g' - kv^2$. This indicates that the skier has a limiting velocity, $V_T = (g'/k)^{1/2}$, and, since $V_T = 55 \, \mathrm{m s^{-1}}$ for $\theta = 40°$, $k = 0.0018 \, \mathrm{m^{-1}}$. If this k-value also applied to the skier on the 15° slope, he would have $V_T(15°) = (1.59/0.0018)^{1/2} = 29.7 \, \mathrm{m s^{-1}}$. Even this could well be an overestimate, because a record breaking skier would prepare his skis carefully to reduce the coefficient of friction by as much as a factor of two. If that were the case, k would be increased to 0.002 and the terminal velocity on the 15° slope would be $28.5 \, \mathrm{m s^{-1}}$.

By using the same value of k in both the steep and shallow slope cases, we have assumed implicitly that the skiers present the same area of cross-section perpendicular to their direction of travel. It seems likely, however, that the high-speed skier will crouch as far as possible to reduce this area, while the skier on the shallow slope might go so far as to remain upright. This might roughly double the value of k for the slower skier relative to the expert. Then V_T for the 15° slope would be reduced to around $25 \, \mathrm{m s^{-1}}$.

The relationship of k to cross-sectional area can be established by referring to the argument leading to the proposition that $F_r \propto v^2$. There we established the plausibility of a force $F_r = \rho A v^2$, but we emphasized that it was only roughly appropriate. If, for example, we say that the momentum loss on impact of the elements composing the fluid can range from v to 0 per unit mass, and that all changes are equally probable, then the average specific momentum change is $v/2$ and the resistive force is halved. In fact, we have little idea in general of the distribution of momentum exchange, except that we might say that it depends on the shape of the body. This leads to the possibility of using an equation of the form $F_r = \frac{1}{2}\rho C_D A v^2$, where C_D is a (dimensionless) coefficient of drag which depends on the details of the momentum exchange between the fluid and the body moving through it. C_D depends on the shape of the body, of course, and is an empirical correcting factor.

8.2. The car makes progress because its tyres push against the road surface. This limits the force available to drive the vehicle forwards, because the maximum force which the tyres can exert in propulsion is the contact force between the tyres and the road surface. We can write this frictional force in terms of the normal force, $F_m = \mu N$, and if, for simplicity, we take $N = mg$, $F_m = 0.6 \times 9.8 \times 1000 = 5.88$ kN. The force at the wheels due to the power supplied from the engine may be estimated as $F = P/v$. For small values of v, F will be relatively large and may exceed F_m, producing wheelspin. This situation will continue until $F = F_m$; that is, $F_m = P/v$. The maximum power will become available when $v = P/F_m = 120/5.88$, or $v = 20.4$ m s^{-1}. For velocities greater than this, the power will be converted all to driving force, which will become smaller in inverse proportion to the velocity.

The progress of the car is force limited at low velocities and power limited at higher velocities. We can find the distance travelled in the force limited range because the force driving the vehicle is constant in this range. We may then say that the work done by the driving force is equal to the kinetic energy of the vehicle, $\frac{1}{2}mv^2 = F_m x$, where x is the distance travelled. This leads to $x = 35.4$ m. The time taken is $t = vm/F_m = 3.5$ s, with an acceleration of 5.8 m s^{-1}.

8.3. The commonly used kph doesn't match our system of units and must be converted to m s^{-1}. Since 1 kph is equivalent to 1000 m per 3600 s, 1 kph $= 0.2778$ m s^{-1} so that 48 kph $= 13.33$ m s^{-1} and 80 kph $= 22.22$ m s^{-1}. The slope rises 1 m vertically in 20 m horizontally, so that its angle is $\theta = \tan^{-1}(1/20) = 2.862°$. At maximum speed uphill the forces acting down the slope are a resistive force, kv^2, and the component of weight parallel to the slope, $mg \sin\theta$. The uphill force is the power of the scooter divided by its velocity, P/v_u. These are equal for the constant velocity $v_u = 13.33$ m s^{-1}; that is,

$$kv_u^2 + mg \sin\theta = P/v_u.$$

For the downhill uniform velocity, $v_d = 22.22$ m s^{-1}, the resistive force and the component of the weight down the slope are opposed, so that

$$kv_d^2 - mg \sin\theta = P/v_d.$$

We have two equations in the two unknown quantities, k and P. On substituting in the values given we find that $k = 0.303$ and $P = 1697$ kW. For motion in the horizontal plane, $kv^2 = P/v$, or $v = (P/k)^{1/3}$; that is, $v = 17.75$ m s^{-1} or 63.9 kph.

8.4. Once again we have to convert from kph to m s^{-1}. This gives 96 kph $= 26.67$ m s^{-1} and, at maximum speed the tractive force, $F = P/v = kv^2$, giving $k = P/v^3$. Substituting $P = 7.5 \times 10^5$ leads to $k = 39.55$ kg m^{-1}. The change in the units of k relative to Exercise 8.1 arises because here we have defined the resistive force as kv^2, whereas in Exercise 8.1 it was defined as mkv^2. At a velocity v (< 96 kph) the force acting on the train accelerates it, so that the equation of motion is

$$mv \frac{dv}{dx} = \frac{P}{v} - kv^2.$$

Multiplying through by v and integrating gives

$$m \int \frac{v^2 \, dv}{P - kv^2} = \int dx.$$

The integral in v is simple because its form is analogous to the integral in Example 35. This time, the derivative $d(kv^3) = 3kv^2 \, dv$, so that

$$-\frac{m}{3k} |\ln(P - kv^3)|_0^{13.3} = X$$

$$= \frac{m}{3k} \ln\left(\frac{P}{[P - k(13.3)^3]}\right).$$

Substituting $k = 39.55$, $P = 7.5 \times 10^5$, and $m = 3 \times 10^3$ kg gives $X = 337.6$ m.

8.5. The force acting downwards is the weight of the parachutist, and the dissipative force acts upwards. This time the acceleration equation, in terms of the distance fallen, y, is

$$v \frac{dv}{dy} = g - kv^2,$$

so that $V_T^2 = g/k$.

We may write $g - kv^2 = (g/V_T^2)(V_T^2 - v^2)$, so that on separating the variables and integrating we have

$$\int \frac{v \, dv}{V_T^2 - v^2} = \frac{g}{V_T^2} \int dy,$$

so that

$$-\tfrac{1}{2} \ln(V_T^2 - v^2) = (g/V_T^2)y + C.$$

Since $v = 0$ when $y = 0$, $C = -\tfrac{1}{2} \ln(V_T^2)$. Then

$$[1 - (v^2/V_T^2)] = \exp(-2gy/V_T^2)$$

or

$$v = V_T[1 - \exp(-2gy/V_T^2)]^{1/2}.$$

The velocity will be within 1% of V_T when the square bracket has the value of 0.99; that is,

$$\exp(-2gy/V_T^2) = 0.01, \quad \text{or} \quad [(2 \times 9.8)/55^2]y = \ln(100),$$

leading to $y = 660$ m. The parachutist's altitude is then 2340 m.

In unresisted fall the parachutist's velocity when $y = 660$ m would be $v = (2gy)^{1/2}$, which is about $114 \, \text{m} \, \text{s}^{-1}$, well over double his terminal velocity.

9
Variable forces

9.1 Variation of *g* with altitude

We have examined the effect of gravitational fields on masses and seen that Coulomb's inverse square law applies in electrostatic problems. Even though the electrical potential and the gravitational potential energy have been defined, we have restricted the range of our interest to the cases in which, for example, the acceleration due to gravity, g, is effectively constant. Since the gravitational force acting on a body of mass m distant r from the centre of a sphere of mass M is $F = GMm/r^2$, with G the universal constant of gravitation, we can form an expression for the weight of the body of mass m as $mg = GMm/r^2$. Cancelling m leaves the statement $g = GM/r^2$, which is strictly constant only for a given value of r. Effectively, g is constant over a range of r, say from r to $r + \delta r$, in which the change in r^2, $\delta(r^2)$, may be neglected. Since the earth is approximately spherical with a radius of 6371 km, the change in g for 1 km change in altitude (measured from the surface of the earth) is only 0.03%, and for many purposes may be neglected. On the other hand, when we are dealing with situations involving large changes in altitude, for example in the ballistics of a rocket, neglecting the variations of g over the path followed by the rocket would lead to a considerable error.

Example 48. A rocket is to be fired so that it will escape from the influence of the earth's gravitational field. What velocity must it have at the surface of the earth so that it can make its escape successfully?

The force acting on the rocket of mass m at the earth's surface will be its weight, $mg = GMm/r^2$, where $G = 6.673 \times 10^{-11}\,\mathrm{N\,m^2\,kg^{-2}}$. The mass of the earth may be taken as 6×10^{24} kg, and its radius R as 6.37×10^6 m. As the rocket makes its progress away from the surface of the earth then, at a distance r from the earth's centre, the force of gravitational attraction is $F(r) = -GMm/r^2$. In travelling the small distance δr, from r to $r + \delta r$, the work done by the rocket will be $-F(r)\delta r$. If the rocket is to escape, its initial kinetic energy must be at least equal to the work it has to do in order to arrive at the value of r for which $F(r) = 0$; that is,

$$-\int_R^\infty F(r)\,\mathrm{d}r = -\int_R^\infty \frac{GMm}{r^2}\,\mathrm{d}r = \frac{GMm}{R} = -V(R),$$

where $V(R)$ is simply the potential energy of the rocket at the earth's surface. The initial kinetic energy of the rocket must be at least equal to the magnitude of its gravitational potential energy, so that $T + V(R) \geq 0$. Thus

$$\tfrac{1}{2}mv^2 \geq GMm/R, \qquad \text{leading to} \qquad v \geq 11.2\,\mathrm{km\,s^{-1}}.$$

This is, of course, a simplistic model because, initially, the rocket will be at rest on its launch pad and will be accelerated from rest by its motors. If it were a single stage rocket designed so that it reached its maximum velocity at an altitude of 100 km, the velocity required for it to escape would be that for which $v^2 = 2GM/(6.47 \times 10^6)$, that is, $v = 11.1 \, \text{km s}^{-1}$, an almost negligible change from the previous case. At an altitude of 1000 km the escape velocity has been reduced only to $10.4 \, \text{km s}^{-1}$, while to halve the escape velocity the 'launch' must take place at a distance of $2R$ from the earth's centre, at an altitude of 6379 km. The escape velocity that we have calculated for the rocket situated initially on the earth's surface provides us with an upper limit for the escape velocity. Our neglect of dissipative effects—that is, of the resistance by the air to the rocket's motion—means that we have underestimated the energy required to drive the rocket upwards. Fortunately, the atmosphere becomes less dense quite rapidly with height above the earth's surface, so that at an altitude of a few tens of kilometres its effects become relatively small.

We could have approached this problem by observing that when the rocket has its minimum escape velocity it will have zero velocity when it arrives at a point infinitely distant from the earth's surface. Its total energy at that point will be zero. Energy conservation requires that its total energy at the earth's surface is zero; that is, $\frac{1}{2}mv^2 - GMm/R = 0$, just as before. This problem of escape velocity is closely related to the energy considerations about the bound state with which we met in Example 34.

9.2 Variable mass

There is another aspect of rocket mechanics which we have neglected so far. It is a matter of observation, of pyrotechnic rockets or of space vehicles, that the rocket carries a large mass of fuel for its propulsion and the fuel is consumed as the rocket progresses. The mass which is accelerated when a rocket is launched varies from that of the rocket shell and most of the fuel it contains to that of the shell only. A second stage in our approach to rockets must take this variation of mass into account. Generally speaking, the propulsive gases have a particular (fixed) velocity with respect to the rocket shell, say $-v_0$, negative because we choose the direction of the rocket's velocity as positive. When the rocket alters its mass from m to $m - \delta m$ by ejecting its burning fuel while its velocity increases from v to $v + \delta v$, there is no change in the combined momentum of the rocket and the exhaust gases because there is no external force acting on the rocket–fuel system. This conservation of momentum may be written in terms of the changes in momentum of the rocket, $m\delta v = (m - \delta m)\delta v$, and of the fuel, $(-\delta m)v_0$, as

$$m\delta v - v_0 \delta m = 0$$

or

$$\delta v = -(\delta m/m)v_0$$

which, in the limit as $\delta m \rightarrow 0$ becomes

$$dv/dm = -v_0/m.$$

If the rocket accelerates from an initial velocity v_i to a final velocity v_f while its mass reduces from m_i to m_f, we may integrate this equation as

$$\int_{v_i}^{v_f} dv = -v_0 \int_{m_i}^{m_f} \frac{dm}{m}$$

to give

$$v_f - v_i = -v_0 \ln(m_f/m_i).$$

which is the same as

$$v_f - v_i = v_0 \ln(m_i/m_f).$$

The initial mass is bound to be greater than the final mass, so that the change in velocity is an increase in velocity, just as we would expect.

If we are interested in the acceleration of the rocket or the thrust of the exhaust gases we may write

$$\frac{dv}{dm} = \left(\frac{dv}{dt} \right) \left(\frac{dt}{dm} \right),$$

so that

$$\frac{dv}{dt} = -\frac{v_0}{m} \frac{dm}{dt}$$

or

$$\text{thrust} = m \frac{dv}{dt} = -v_0 \frac{dm}{dt}.$$

This equation has to be modified when there are gravitational forces acting on the rocket. In the case of a rocket launched in a vertical direction with respect to the earth's surface, this is a simple matter of subtracting the rocket's weight from the right-hand side of the equation, to give

$$m \frac{dv}{dt} = -mg - v_0 \frac{dm}{dt}.$$

This scalar equation for the vertically launched rocket may be integrated so that, if the initial velocity is 0, the final velocity is v_f, and the mass changes from m_i to m_f,

$$\int_0^{v_f} \frac{dv}{dt} dv = -g \int_0^t dt - v_0 \int_0^{m_f} \frac{1}{m} \frac{dm}{dt}$$

which gives

$$v_f = -gt + v_0 \ln(m_i/m_f).$$

When the velocity of the rocket is in a direction which is not radially away from a centre of gravitational attraction, this equation is again modified by substituting vectors for the scalar v's and g, leading to

$$m \frac{dv}{dt} = mg + \frac{v_0}{m} \left(\frac{dm}{dt} \right).$$

Example 49. A rocket of mass 60 tonnes is set on its launch pad for vertical firing. Its exhaust gases are delivered with a velocity of $3 \, \text{km s}^{-1}$ relative to the rocket shell. What is the rate at which gas must be ejected from the rocket (i) to counteract the weight of the rocket, and (ii) to give the rocket an initial acceleration of $4g$? If the rocket carries 40 tonnes of fuel, what will be its highest velocity, assuming that the rate of discharge of the fuel is maintained constant and that air resistance can be neglected?

The equation of motion gives the equation for acceleration as

$$\frac{dv}{dt} = -g + \frac{v_0}{m}\left(\frac{dm}{dt}\right).$$

The acceleration will be zero when the thrust of the rocket counteracts the weight of the rocket. When this is the case, $dm/dt = mg/v_0$, so that the discharge rate is $196 \, \text{kg s}^{-1}$.

For an acceleration of $4g$ we have $dm/dt = 5mg/v_0$, giving a discharge rate of $981 \, \text{kg s}^{-1}$, very nearly one tonne per second.

When all the fuel has been burned, the rocket will have attained its final velocity. If we identify the discharge rate as $1 \, \text{tonne s}^{-1}$ this will take approximately $40 \, \text{s}$. Then the velocity of the rocket will be

$$v = -40g + [3 \times 10^3 \ln(6/2)] = 2896 \, \text{m s}^{-1}.$$

What is noticeable about this result is that without the gravitational attraction term the velocity of the rocket shell would have been more than that of the exhaust gases relative to the shell. As example, in the absence of gravitational forces a value of m_i/m_f of 7.4 would give a final velocity $v \simeq 2v_0$. This apparent 'enhancement' of the velocity arises even though the force (the thrust of the rocket engines) remains constant. This is because the constant force acts on a diminishing mass, so that the acceleration is not constant but an increasing function of time. This observation puts a limit on the allowable rate of discharge of the exhaust gases, because the payload of the rocket is unlikely to be able to withstand very large accelerations. The maximum acceleration with the constant rate of discharge, $1 \, \text{tonne s}^{-1}$, occurs when the fuel is almost all consumed and the rocket mass is effectively the mass of the shell alone. The acceleration then is about $16g$, most likely an intolerably high value. Alternative strategies to avoid this problem are either to vary dm/dt as the rocket progresses or to reduce the initial acceleration to the extent that the largest acceleration is less than the acceptable limit. If this limit were $10g$ the constant discharge rate would need to be halved (roughly) from the value calculated above. This would reduce the initial acceleration to about $2g$. On the other hand, the discharge rate could be made a decreasing function of time so that the acceleration didn't reach its allowed limit.

If the 60 tonne rocket was constructed in two stages with masses of 45 and 15 tonnes carrying 30 and 10 tonnes of fuel respectively, the first stage would increase the velocity of the combination from zero to about $1800 \, \text{m s}^{-1}$. The second stage would then increase its velocity from $1800 \, \text{m s}^{-1}$ to about $3800 \, \text{m s}^{-1}$, nearly twice the final velocity of the single stage rocket. This is a much more effective use of the fuel.

Once more, we have neglected the effect of air resistance. At lower altitudes fuel will be consumed to provide the energy (or do the work) to overcome this dissipative force.

9.3 The linear oscillator

In the two previous examples we have dealt with a force which is variable as a function of position and of a constant force acting on a variable mass. A second example of a position variable force (which we have examined previously) is the force exerted by a stretched (or compressed) spring. This force is directed so that it tends to restore the spring to its equilibrium position. Take the situation in which a body of mass m hangs on the end of a vertical spring of spring constant k. When the system is in equilibrium the weight of the body is equal to the tension in the spring, $mg = kh$, where h is the extension produced in the spring's length by the mass suspended from it. What happens if the mass is given a vertical displacement, x, from its equilibrium position? The tension in the spring becomes $k(h + x)$, so that the force tending to restore the body to its equilibrium position is $[kh - k(h + x)] = -kx$. The acceleration suffered by the body is then

$$\frac{d^2x}{dt^2} = -\frac{kx}{m}$$

or, since

$$\frac{d^2x}{dt^2} = v\left(\frac{dv}{dx}\right),$$

this may be written as

$$v\left(\frac{dv}{dx}\right) = -\frac{kx}{m}.$$

On integrating, we have

$$\tfrac{1}{2}v^2 = -\tfrac{1}{2}\frac{kx^2}{m} + C$$

so that, on taking $v = 0$ when $x = a$ (the maximum displacement), we find that

$$v^2 = (k/m)(a^2 - x^2),$$

or, setting $k/m = \omega^2$,

$$v = \omega(a^2 - x^2)^{1/2}.$$

Since v is the same as dx/dt, this equation giving v in terms of x is also an equation giving x in terms of t:

$$dx/dt = \omega(a^2 - x^2)^{1/2}.$$

This equation may be integrated to give

$$\sin^{-1}(x/a) = \omega t + C',$$

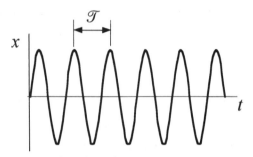

Fig. 9.1 Displacement, x, as a function of time, t, for the oscillating mass supported by a spring. The period, $\mathcal{T}$, of oscillation, is shown.

so that if $x = a$ when $t = 0$, $C' = 90°$ and $x = a \sin(\omega t + 90°) = a \cos \omega t$. If we can write $x = a \cos \omega t$, then $v = dx/dt = -a\omega \sin \omega t$, and the acceleration is $-\omega^2 a \cos \omega t = -\omega^2 x$, demonstrating that the sinusoidal trignometric function provides a solution which is consistent with the equation of motion. A plot of x as a function of t is a simple sinusoidal function, the initial value of which is determined by the boundary condition: x returns to its initial value at the time $\mathcal{T}$ at which the sinusoidal function returns to its initial value, as shown in Fig. 9.1. Since this happens for $\omega t = 2\pi, 4\pi$, and so on, we can relate ω to $\mathcal{T}$ by $\mathcal{T} = 2\pi/\omega$. $\mathcal{T}$ is known as the period of oscillation and ω as the angular frequency of the oscillator. The maximum displacement, a, is called the amplitude of the oscillations. Because the force producing these oscillations is linear in displacement, the oscillations are those of a linear oscillator. Alternatively, because the oscillations can be described by the simple sinusoidal function, the oscillations are called simple harmonic oscillations and the oscillator is known as a simple harmonic oscillator.

Example 50.   The scale pan on a culinary spring balance (kitchen scales) has a mass of 20 g. When placed on the scales it compresses the spring by 2 mm. What would the period of oscillation of the scale pan be if it were displaced slightly from this equilibrium position?

The static deflection of the spring is related to the force producing it by $F = -kx_0 = mg$. This gives $k = mg/x_0 = 98 \, \text{N m}^{-1}$. After the displacement from this equilibrium position, the force acting on the scale pan is $m(dv/dt) = -mgx/x_0$. The angular frequency, ω, is then $\omega = (g/x_0)^{1/2}$, so that the period of oscillation is $\mathcal{T} = 2(x_0/g)^{1/2}$, which has the value 0.09 s or 90 ms. The number of oscillations in a second, the frequency of oscillation, $n = 1/\mathcal{T} = 11.1 \, \text{Hz}$.

A point to be noticed here is that the mass of the scale pan does not enter into the expression for the frequencies and the period of the oscillations. This appears to contradict our initial derivation of the expressions for the displacement, velocity, and so on, for the linear oscillator. Although m does not appear explicitly in the frequency, it is there implicitly in the static deflection x_0.

It's easy to find an expression for the energy of a linear oscillator. For the velocity, v, of simple harmonic oscillations of amplitude a we have $v = \omega(a^2 - x^2)^{1/2}$, where ω is the angular frequency of the oscillations. On squaring and multiplying by half the oscillator mass, we have

$$\tfrac{1}{2}m\omega^2 a^2 = \tfrac{1}{2}mv^2 + \tfrac{1}{2}m\omega^2 x^2.$$

The mass m seems to appear in this equation only because we used it as a multiplier, but ω is not independent of m. If the oscillator has a spring constant k, then $\omega^2 = k/m$ and the expression for the energy becomes

$$\tfrac{1}{2}ka^2 = \tfrac{1}{2}mv^2 + \tfrac{1}{2}kx^2.$$

The second term on the right-hand side of this equation, and the left-hand side term, are seen to be the same as the expressions for the potential energy associated with the compressions or extensions, x or a, of a spring of spring constant k.

The total energy of the system is constant (at the value $\tfrac{1}{2}ka^2$), just as we would expect, and the oscillations represent the continual conversion of potential energy to kinetic energy and vice versa. Had we begun our examination of the linear oscillator by writing down an expression for the total energy of the system, we could have obtained the expression for v without using the integration process. The most important point about the energy equation for the linear oscillator is that it tells us that the total energy of the oscillator is proportional to the square of the amplitude of the oscillation, a^2, to which it is subjected. In terms that we have used previously,

$$T + V = H, \qquad H \propto a^2.$$

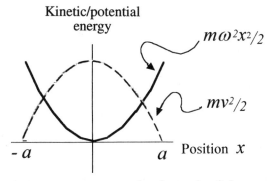

Fig. 9.2 The kinetic and potential energies, $\tfrac{1}{2}mv^2$ and $\tfrac{1}{2}mx^2\omega^2$ respectively, for a linear oscillator of mass m, of angular frequency ω, and of amplitude a, as functions of its displacement, x, from the centre of its path. The maximum value attained by either of these energies is $\tfrac{1}{2}ma^2\omega^2$.

Example 51.   The vertical vibrations of a road vehicle's body may be modelled most simply by assuming that it is a mass which rides on vertical springs. A bus of mass 2 tonnes can carry 35 persons (including the driver) whose average mass is 60 kg. If the bus sinks on its springs by 5 cm when the driver and passengers enter to fill it completely, what will be the frequency of its vertical oscillations when it is fully laden? What would the corresponding frequency be if the bus carried only the driver?

The static deflection of the springs as the bus fills gives a measure of the spring constant, since $kx = mg$: $k = (2100 \times 9.8)/0.05 = 0.41\,\text{MN}\,\text{m}^{-1}$. The equation of motion is $m(dv/dt) = -kx$, so that the angular frequency is $\omega = (k/m)^{1/2}$, where m is the total mass carried by the springs, the mass of the bus, and its contents. The frequency of oscillation is $n = 1/\mathcal{T} = \omega/2\pi$. When the bus is full $n = n_f = 1.59\,\text{Hz}$, and when it is empty of passengers $n = n_e = 2.25\,\text{Hz}$.

Anyone who has given more than a cursory glance at a cutaway drawing of a motor vehicle will realize that we have used the simplest model possible. Nonetheless, it has justified the observation that a bus carrying few passengers will be subjected to vibrations of a higher frequency than it would be if it were loaded to capacity.

Example 52. A piston in the cylinder of a petrol engine has a stroke of 5 cm and executes simple harmonic oscillations. If the piston completes 6000 oscillations in 1 minute, what is the maximum speed with which it moves?

The frequency of oscillation of the piston is $6000/60 = 100\,\text{Hz}$ and its angular frequency, $\omega = 2\pi \times 100 = 628\,\text{s}^{-1}$. The piston speed may be represented by $v = \omega a \sin \omega t = \omega(a^2 - x^2)^{1/2}$, the maximum value of which is $v = \omega a$. Since the amplitude is half the stroke, $a = 0.025\,\text{m}$ and $v_m = 15.7\,\text{m}\,\text{s}^{-1}$.

Example 53. A group of weight watchers decide to devise a precise method of weighing themselves. They fix upon measuring the frequency of vertical oscillation of a chair weighing 6 kg mounted on a spring, and the corresponding frequency when one of their members sits in the chair. If the spring has a spring constant of $588\,\text{N}\,\text{m}^{-1}$, what will be the frequency of oscillation of the unoccupied chair? If the subject of measurement has a mass of 48 kg, what will the frequency of oscillation be when he sits in the chair?

The equation of motion is $m(dv/dt) = -kx$, so that the angular frequency of oscillation is $\omega = (k/m)^{1/2}$. With $m = 6$, $\omega = 9.9$ and the frequency of oscillation is $n = \omega/2\pi = 1.58\,\text{Hz}$. When m increases to 54 kg the new angular frequency is $3.3\,\text{s}^{-1}$ and the new frequency is 0.53 Hz. Alternatively, the period of oscillation is increased from 0.63 s to 1.9 s.

For any one value of m, $m = (k/4\pi^2 n^2)$, where n is the frequency corresponding to a load of m. The precision of measurement needs to be an improvement on the 1% or so expected from bathroom scales to make the weight watchers' efforts worthwhile.

Example 54. While assisting in the kitchen, a small child notices that a vertical displacement of the kitchen scales scale pan causes it to perform oscillations. If these oscillations were simple harmonic with a maximum displacement of 1.2 cm, and had a period of oscillation of 3 s, what would be the maximum values of velocity and acceleration that the scale pan would attain? When the pan reaches its position of maximum upward displacement the child, hoping to increase the amplitude of the motion, strikes it a sharp blow in the downward direction. If the mass of the moving part of the scales is 60 g and the amplitude of the motion after the blow is 1.5 cm, what impulse has the child delivered to the scales?

The period of oscillation, $\mathcal{T} = 2\pi/\omega$ gives the angular frequency $\omega = 2\pi/3$. If the displacement of the pan at a time t is $y = a \sin \omega t$ (with a the amplitude of the motion), then $v = a\omega \cos \omega t$ and the acceleration $dv/dt = -\omega^2 a \sin \omega t$. The maximum values of these quantities occur when the sinusoidal functions $\cos \omega t$ and $\sin \omega t$ have the value 1. Thus

$$v_m = a\omega = (2\pi/3) \times 0.012 = 0.025 \, \mathrm{m\,s}^{-1} = 2.5 \, \mathrm{cm\,s}^{-1}$$

and

$$(dv/dt)_m = \omega^2 a = 5.26 \, \mathrm{cm\,s}^{-2}.$$

The impulse, J, of the downward blow will give the scale pan a velocity of $J/m = 16.67 \, \mathrm{J\,m\,s}^{-1}$. This velocity can be described in terms of the position of the scale pan by

$$v^2 = \omega^2 (A^2 - y^2),$$

where A is the new amplitude of the motion, 1.5 cm, and y is the displacement when the impulse is given. On substituting the values of m, A, and y, we obtain $J = 0.011 \, \mathrm{N\,s}$.

Here the relationship between J and v^2 is particularly simple, because the increase in momentum is from a state of zero velocity. Had the impulse been delivered at some arbitrary position y, then the new velocity would be $v(y) + (J/m)$ and the equation to be solved would be of the form

$$[v(y) + (J/m)]^2 = \omega^2 (A^2 - y^2)$$

or

$$J/m = \omega \left[(A^2 - y^2)^{1/2} - (a^2 - y^2)^{1/2} \right],$$

so that both a and A need to be known to solve for J.

9.4 Addition of harmonic motions

We are all familiar with the conversion of linear motion to rotational motion. The most commonplace example is in piston-engined vehicles, in which the change from the reciprocating motion of the pistons to rotation of the driving wheels provides the means to drive the vehicle. This change, by means of the crankshaft, may be modelled as the superposition of two simple harmonic motions, with the same frequencies and amplitudes, the directions of motion of which are at right angles to one another, say in the x- and y-directions. The x- and y-displacements at a time t of the particle subject to these are

$$x = a \cos \omega t, \qquad y = a \cos(\omega t + \phi),$$

where a is the amplitude and ω is the angular frequency of the linear oscillations: ϕ is the phase difference between the x- and y-oscillations and relates the initial x- and y-displacements when $t = 0$. For convenience, choose $\phi = \frac{1}{2}\pi$, so that when $t = 0$, $x = 0$ and $y = a$. Then

$$x = a \cos \omega t, \qquad y = a \cos(\omega t + \tfrac{1}{2}\pi) = a \sin \omega t$$

and

$$x^2 + y^2 = a^2,$$

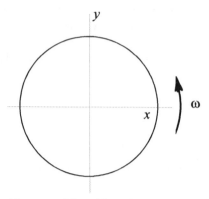

Fig. 9.3 The path followed by a particle subjected to two simple harmonic oscillations of the same amplitude and frequency, with a phase difference of $\frac{1}{2}\pi$, at right angles.

which is an equation describing a circle of radius a, centred at the origin of the xy-plane, as shown in Fig. 9.3. The path of the particle subjected to these simultaneous harmonic motions is a circle, the radius of which is the amplitude of the component oscillations. Since the particle returns to its initial position when the oscillations have completed one cycle of oscillation, this occurs for $\mathcal{T} = 2\pi/\omega$. Then we have $\omega \mathcal{T} = 2\pi$, indicating that the angular frequency associated with the linear oscillations of the particle is the same as the angular velocity of the particle in its circular path.

The circular path obtained in the case of a phase difference of $\frac{1}{2}\pi$ is nearly the simplest solution to the problem of adding together two linear oscillations in directions which are perpendicular to each other. The simplest solution comes if we choose the initial conditions

$$x = a \quad \text{and} \quad y = \pm a \quad \text{for} \quad t = 0.$$

Then

$$x = a \cos \omega t$$

and

$$y = \pm a \cos \omega t,$$

so that $x = \pm y$. The sign may be positive or negative because the x- and y-velocities may be of the same sign or of opposite sign. For phase differences other than 0 or $\frac{1}{2}\pi$, the path followed by the particle will be elliptical, with the axes of the ellipse tilted away from the x- and y-axes, even though the amplitudes and frequencies of the two oscillations are the same.

When the two component motions have different amplitudes, a and b, and the same frequency, $x = \pm (a/b)y$ for zero and π phase difference. In all other cases the resulting path of the particle is elliptical, even when the phase difference is $\frac{1}{2}\pi$ or $\frac{3}{2}\pi$. If the frequencies of the two components are different, the path followed by the particle is, in general, complicated. Only when the two angular frequencies, ω_x and ω_y, are integer multiples of each other does a relatively straightforward repetitive path emerge. These paths, simple and complicated, are called Lissajous figures.

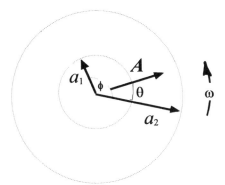

Fig. 9.4 The addition of two vectors, a_1 and a_2, with phase difference ϕ, rotating with the same angular velocity, ω. The result is a vector of amplitude A and phase difference θ from a_1.

A body subjected to the superposition of two simple harmonic motions at right angles, with the same amplitudes, a, and angular frequencies, ω, moves around a circular path of radius a with an angular velocity ω. This observation may be inverted, so that we might say that a body travelling with uniform angular velocity around a circle is subject to two simple harmonic motions of equal amplitude and frequency in directions at right angles to one another. One of those simple harmonic motions can be represented by the projection of the radius vector on to the axis of that motion. This provides us with the possibility of representing a simple harmonic motion by a vector which rotates with an angular velocity equal to the angular frequency of the harmonic motion, and the length of which is the amplitude of the motion. Such a representation is used often to introduce the ideas and formulas of simple harmonic motion. Its advantage, though, is that it provides us with the possibility of forming the sum of simple harmonic motions along the same direction simply by adding the rotating vectors which represent them. Let's say that we have two such motions with the same frequency but with different amplitudes, a_1 and a_2. The vectors representing these are rotating with the same angular velocity, ω, so that the angle between their directions, their phase difference, ϕ, will remain constant with time. Simply summing the two vectors, as illustrated in Fig. 9.4, provides us with the resultant amplitude (from the cosine rule) as

$$A^2 = a_1^2 + a_2^2 + 2a_1a_2 \cos \phi,$$

so that for $\phi = 0, 2\pi, \ldots$, for which $\cos \phi = 1$,

$$A^2 = (a_1 + a_2)^2,$$

while for $\phi = \pi, 3\pi, \ldots$,

$$A^2 = (a_1 - a_2)^2,$$

just as we would expect. The phase difference between the vectors A and a_1 is obtained from the sine rule, as $\theta = \sin^{-1}[(a_2/A)\sin \phi]$.

A body subjected to two simple harmonic motions of the same frequency with different amplitudes, and with a phase difference, ϕ, will perform simple harmonic oscillations with the same frequency but with a different amplitude, A, and a phase difference from the component motions.

9.5 Amplitude modulation

When the frequencies of the two component harmonic motions are different, the result of their superposition is more complicated, but may be illustrated by saying that the two components have the form

$$x_1 = a \cos \omega_1 t \quad \text{and} \quad x_2 = a \cos \omega_2 t$$

and that they have the same amplitudes and are in phase ($\phi = 0$) when $t = 0$. As t increases, the two components will go out of phase because the vectors representing the motions have different angular velocities. The phase difference will increase with time until the vector rotating at higher angular velocity 'catches up' with the other one. The two component motions will then be in phase again. This process will then repeat itself, so that the amplitude, A, of the resultant motion will be a varying function of time. This qualitative result can be made quantitative by adding together the two x's to give a resultant displacement, X, as

$$X = a[\cos \omega_1 t + \cos \omega_2 t],$$

which is identical to

$$X = 2a[\cos\{\tfrac{1}{2}(\omega_1 - \omega_2)t\}][\cos\{\tfrac{1}{2}(\omega_1 + \omega_2)t\}].$$

This equation may be interpreted as the product of the second sinusoidal term and an amplitude which varies with time such that

$$A(t) = 2a[\cos\{\tfrac{1}{2}(\omega_1 - \omega_2)t\}],$$

so that the resulting motion is a harmonic motion (no longer simple, however) with angular frequency $[\tfrac{1}{2}(\omega_1 + \omega_2)]$, the amplitude of which varies from zero to $2a$ and back to zero in a time period $\mathcal{T}_A = [2\pi/(\omega_1 - \omega_2)]$.

The form of this 'combination' harmonic motion is shown in Fig. 9.5. It should be remembered that the reality of the situation is given by the form of the wave

Fig. 9.5 The result of adding together two simple harmonic motions along a line with different frequencies. The amplitude of the resulting motion varies with time.

shape inside the envelope joining successive points of maximum or of minimum on the curve. The envelope modulates the amplitude of that harmonic motion, imposing a sinusoidal variation of its amplitude with time.

9.6 Waves

This process of adding together two simple harmonic motions along a line or perpendicular to each other may be extended to explain some phenomena associated with wave motion. A wave represents the transmission of energy through a solid, liquid, or gas (a medium) by means of some form of coupling together between adjacent elements of the medium. When one of the elements is set into motion, its neighbours tend to move as well. If a boundary element of a medium is induced to perform simple harmonic oscillations, the oscillation will progress through the medium. Energy will be transferred by this means to the other boundaries of the medium, transmitted by a wave. When the source that drives the simple harmonic motion of the boundary element, the 'end' of the wave progresses through the medium to the other boundaries and the medium returns to its initial state. Not surprisingly, for a simple wave with a sinusoidal wave profile, the rate at which energy is transferred by a wave is proportional to the square of its amplitude, which is the same as the amplitude of the simple harmonic motions executed by the elements of the medium in the path of the wave. This identification of wave motion essentially as a succession of simple harmonic oscillations leads to the idea that a transverse wave (in which the oscillations are perpendicular to the direction in which the wave travels), induced by simple harmonic motions which are at right angles to each other, is one in which the elements of the medium may execute linear oscillations, or travel around elliptical, or circular, paths. These more complicated motions of the elements of the medium as the wave passes enable us to visualize a transverse wave that is linearly, or elliptically, or circularly, polarized, respectively. When the two simple harmonic motions that drive the boundary element to produce the wave are along the same line and have the same amplitude and different frequencies, the harmonic motion of the elements of the medium in the path of the wave will have a modulated form, as shown in Fig. 9.5. The wave profile will have a variable amplitude, and energy will be transmitted through the medium by an amplitude modulated wave. This energy will arrive at a detector with an intensity that varies with the angular frequency $\frac{1}{2}(\omega_1 - \omega_2)$. This form of modulation gives an explanation of the beating of sound waves to produce a signal lower in frequency than its component waves. It will also be louder than either of its component waves.

Example 55. In bungy jumping, the jumper dives from a platform at a considerable height above ground level towards the ground below. He is attached to one end of an elastic rope, the other end of which is anchored on the platform. His survival depends on the properties of the rope. This must be so elastic that it does not overstress his body, but not so elastic that his fall is terminated by collision with the ground. Say that the jumper has a mass of 90 kg and chooses a 30 m rope which will extend to twice its length when a mass of 900 kg is hung from it. What is the minimum height from which he can safely dive? In the absence of dissipative forces, what would be the period of his oscillation at the end of the rope?

The spring constant can be obtained from the rope specification. A weight of $900g$ N doubles its length so that $900g = k(60 - 30)$, or $k = 30g$ N m^{-1}. This means that the equilibrium position of the jumper when he simply hangs from the end of the rope is $x = 90g/k = 3$ m below the position of the rope end if it were unloaded. We can say that the static extension is 3 m and that this equilibrium position is 33 m below the diving platform.

The length of fall comes from energy considerations. Let the length of the rope be l, the mass of the jumper be m, the static extension be x, and the distance of fall below the equilibrium position be d. Then, equating potential energies, we have

$$mg(d + x + l) = \tfrac{1}{2}k(x + d)^2.$$

Since we know that $x = l/10$ and $k = 900g/l$, we obtain

$$90g[d + (11l/10)] = [(l/10) + d]^2 \times (900g/l),$$

the solution of which is $d = l(21/100)^{1/2}$ which, with $l = 30$, leads to $d = 13.75$ m. The fall takes place over a distance of 46.75 m, which is the minimum height for the platform. The maximum acceleration to which the jumper is subjected should be known in order to ensure his safety. The acceleration will be a maximum when the rope is at its maximum extension, and so we need to know the equation of motion of the diver as the rope stretches. This is simply

$$m(\mathrm{d}^2y/\mathrm{d}t^2) = -ky,$$

where y is the displacement of the diver below 33 m. (The derivation of this equation is the same calculation as if we had hung a weight on a spring and displaced it vertically.) On substituting the values of m and k we find that the angular frequency of the motion is $\omega = (g/3)^{1/2}$. The maximum acceleration occurs for the maximum value of y, 13.75, so that its value is $(13.75g)/3 = 4.5g$ m s^{-2}. If the harness at the rope end is properly designed, this acceleration is unlikely to do permanent damage to the diver, even though a lower value might be more acceptable. Allowing a margin of error because the rope might be below specification leads to the conclusion that the jump could be made safely from 50 m or more provided that the harness design is suitable.

The period of oscillation of the diver if there were no energy losses due to dissipative forces (of his undamped motion) is a combination of the harmonic oscillation while the rope is stretched and the motion under the influence of gravity when the rope is slack. For the 'non-elastic' motion, the time taken to fall 30 m from the platform (or to rise through the last 30 m to the platform) is given by $l = \tfrac{1}{2}gt_1^2$, or $t_1 = (60/g)^{1/2}$. For one 'up and down' movement of this type, the time taken is $2t_1$ or $4(15/g)^{1/2}$. If the rope did not become slack, the period of oscillation would be $\mathcal{T} = 2\pi/\omega = 2\pi(3/g)^{1/2}$. This oscillation is not complete because the rope goes slack when the diver is 3 m above the equilibrium position. The time which would have been taken to travel from the maximum height (at height d above the equilibrium position) had the rope not gone slack to this position may be obtained from the expression for the displacement, $x = d \cos \omega t_2$, which gives $3 = 13.75 \cos \omega t_2$, or $\cos \omega t_2 = 0.22$, leading to $t_2 = [(\cos^{-1}\{0.22\})/\omega]$. The 'missing part of the oscillation' would have taken twice this time, and so the time taken by the elastic oscillation is $[2\pi - 2\cos^{-1}(0.22)][(3/g)^{1/2}]$. When this is combined with the time, $2t_1$ for the 'non-elastic' oscillation, we obtain the period of oscillation as

$$\mathcal{T}_e = 2[\pi + 2(5)^{1/2} - \cos^{-1}(0.22)]\left[(3/g)^{1/2}\right]$$

and, on substituting values, $\mathcal{T}_e = 6.93$ s.

One point which is relevant here is that elastic materials retain their elastic properties only up a certain limiting extension, the elastic limit. For extensions less than the elastic limit Hooke's Law, $F = -kx$, relates the extension of the material to the load it carries. Generally, beyond the elastic limit, the material deforms permanently (plastic deformation) for relatively small increases in the load, and soon fails catastrophically by breaking. When materials are subjected to strains (increases or decreases in length) care must be taken to avoid exceeding the elastic limit; unless, of course, the elastic material is being used as a safety device which will break when a particular value of its load is exceeded. This second use is taking advantage of the 'failure' properties of the material rather than its elastic properties.

9.7 The damped linear oscillator

The calculation of the period of oscillation for the jumper in Example 55 is clearly a little far away from reality. Bungy jumpers do not continue to oscillate at the end of their ropes for any extended period of time. Their oscillations at the end of the rope die down very quickly, perhaps in two or three relatively small up and down movements. The mechanical energy of the jumper–rope system must be dissipated quite rapidly. In the example, the mechanical energy was about 41.7 kJ, so that the dissipative mechanism which transforms the mechanical energy to some other form of energy has to work quite effectively. We have dealt with dissipative forces previously by assuming that the force of resistance to the motion of a body through a fluid is proportional to the velocity of the body at lower rates of movement through the fluid, or to the square of the velocity of the body at higher velocities. It seems quite plausible to include one of these dissipative forces in the equation of simple harmonic motion, to give an equation of simple harmonic motion in which the mechanical energy of the motion is transformed to some other form. This is the equation of damped harmonic motion, in which the oscillations are damped down by the dissipative force. Take the dissipative force as a visco-elastic damping, proportional to the velocity of the body subjected to the motion, $F_r = bv = b(\mathrm{d}x/\mathrm{d}t)$. Say that the mass of the body is m, and the force giving rise to the harmonic motion has the form $F_h = -kx$. Inclusion of the dissipative term modifies the equation of motion from

$$m\frac{d^2x}{dt^2} + kx = 0$$

to

$$m\frac{d^2x}{dt^2} + b\frac{dx}{dt} + kx = 0,$$

or

$$\frac{d^2x}{dt^2} + p\frac{dx}{dt} + \omega_0^2 x = 0,$$

with $p = b/m$ and $\omega_0^2 = k/m$. One solution of this equation which is of considerable interest is for the case of lightly damped motion, when $\omega_0^2 > p^2/4$. This solution has the form

$$x = A[\exp(-pt/2)]\left[\cos\left\{t(\omega_0^2 - p^2/4)^{1/2} + \phi\right\}\right],$$

where ϕ is a phase angle and A is a constant. The displacement, x, is seen to be the product of A, a constant, an exponentially decaying term, $[\exp(-pt/2)]$, and a sinusoidal term, $\cos\{\omega t + \phi\}$, with $\omega = [(\omega_0^2 - p^2/4)^{1/2}]$. The constant has no effect on the general form of x, but the exponential term reduces the value of the sinusoidal term. The initial amplitude of the motion is decreased with time, so that the sinusoidal term diminishes until it becomes effectively zero. This situation is illustrated schematically in Fig. 9.6. The reduction in amplitude is a manifestation of the dissipation of the initial energy of the oscillator—its energy is transformed from mechanical energy to some other form. The rate at which the amplitude decreases is determined by the exponential term, $[\exp(-pt/2)]$. If we measure the amplitudes of successive maximum displacements (separated in time by the period of oscillation, $\mathscr{T} = \{2\pi/\omega\}$) as A_1 and A_2, then the ratio of these

$$\frac{A_1}{A_2} = \frac{\exp(-pt/2)}{\exp(-p\{t+\mathscr{T}\}/2)}$$

$$= \exp(p\mathscr{T}/2).$$

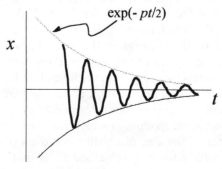

Fig. 9.6 Displacement, x, as a function of time, t, for a lightly damped oscillator. The dashed curves represent the exponential component of x.

In lightly damped systems, for which ω_0^2 is much larger than $p^2/4$, the period of oscillation is unlikely to be significantly different from the period of undamped oscillation. As the damping—the rate of dissipation of energy—is increased, $\omega_0^2 - (p^2/4)$ becomes increasingly different from ω_0^2, producing a marked change in the frequency of oscillation. The period of damped oscillation is then $\mathcal{T} = 2\pi/(\omega_0^2 - \{p^2/4\})^{1/2}$, which tends towards infinity as the value of $p^2/4$ approaches that of ω_0^2. The term 'lightly damped' (or 'underdamped') is used to describe the system in the cases for which oscillatory behaviour is observed; even though, as ω_0^2 and $p^2/4$ become close in value, the damping is not obviously 'light'. When $\omega_0^2 = p^2/4$ the system is 'critically damped' and its amplitude is reduced monotonically from its maximum value to zero. This critical damping provides the quickest possible reduction of amplitude without oscillation.

The lightly damped oscillator loses its mechanical energy under the influence of the dissipative force. When ω and ω_0 are not very different, the rate at which this energy is 'lost' by transformation to other forms of energy is obtained quite easily in terms of the amplitudes, A_1 and A_2, of successive maximum displacements. The potential energies associated with these maximum displacements are

$$V_1 = \tfrac{1}{2}kA_1^2 = \tfrac{1}{2}m\omega_0^2 A_1^2 \quad \text{and} \quad V_2 = \tfrac{1}{2}m\omega_0^2 A_2^2,$$

so that the energy lost by the oscillator in one period of oscillation is

$$\delta H = V_1 - V_2 = \tfrac{1}{2}m\omega_0^2 A_1^2 \left[1 - (A_2/A_1)^2 \right]$$
$$= V_1[1 - \exp(-p\mathcal{T})]$$

or $V_2 = V_1 \exp(-p\mathcal{T})$. This last expression is a solution of the differential equation

$$dH/dt \equiv dV/dt = -pV = -pH = -H/\tau,$$

where $\tau = 1/p$ is a constant time. In a time interval τ the energy of oscillation is reduced by a factor $1/e$, and τ is called the time constant for the energy. The energy of the oscillator is halved in a time of $0.693/\tau = 0.693p$, which could be called the half-life for the energy. It seems obvious that the time constant for the amplitude is $2/p$ and the amplitude half-life is $0.693p/2$.

An alternative way of describing the dissipation of the energy of an underdamped harmonic oscillator is in terms of the quality factor, Q, of the oscillator. This is defined by $Q = \omega\tau = \omega/p$, the number of radians through which oscillation has taken place during the time interval τ. Since Q is proportional to p^{-1}, high values of Q are associated with lightly damped systems.

Exercises

9.1. A plumb line consists of a light string which has a small mass, the plumb bob, attached to one end. The free end of the string is firmly attached to a support so that the string hangs vertically downwards. The string has a length of 0.5 m and the mass of the bob is 50 g. If the bob suffers a small sideways displacement, what is the form of its subsequent motion?

9.2. A simple pendulum with an effective length of 24.8 cm is made from a light string and a spherical bob of 2 cm diameter. It is found that the amplitude of oscillation of the pendulum decreases from 5° to 4.5° in 30 min. If the bob has a mass of 10 g, what is the value of the time constant describing the energy lost by the pendulum due to dissipative forces such as air resistance? Find the time constant for the decay of amplitude and examine how the period of oscillation of the pendulum is affected by the resistive forces.

9.3. The suspension of a motor vehicle may be modelled (as far as vertical oscillations are concerned) as four identical springs symmetrically beneath the bodywork of the vehicle. The vehicle weighs 1 tonne and its bodywork is lowered by 5 cm when four passengers, whose combined weight is 320 kg, get into it. If the weight of the vehicle and its passengers is carried equally by all the springs, by how much are the frequencies of vertical oscillation of one spring different when the vehicle is in this loaded state and in its unloaded state? The vertical motion is moderated by a shock absorber system at each spring. If it is desirable for the amplitude of vibration to be reduced by 63% in one oscillation, what is the value of the damping constant, b (the ratio of the damping force to the velocity), required? What values of b would give rise to critical damping in the loaded and unloaded states?

9.4. The pendulum of a clock oscillates against the combined effects of air resistance and friction. When the clock pendulum is set into free oscillation, the amplitude of oscillation decreases from 5° to 1.84° in 100 s. The pendulum may be represented as equivalent to an underdamped simple pendulum harmonic oscillator of mass 0.2 kg with an undamped period of oscillation of 1 s. At what rate should energy be supplied to the pendulum so that its amplitude of oscillation is maintained at 5°? If the weight which drives the clock has a mass of 2.5 kg and falls through a height of 1.5 m in 7 days, what proportion of the energy supplied to the clock is used to maintain the oscillation of the pendulum?

9.5. A manometer is a device for measuring pressure differences. It consists of a uniform bore glass U-tube with open ends, partially filled with liquid—sometimes mercury, but often oil—usually mounted vertically with the open ends at the top. A difference in the pressures at either end of the tube will cause the fluid to move up the tube in one arm and down the tube in the other. The difference of the heights of the fluid surfaces in the two arms provides a measure of the pressure difference. In such a manometer the total length of the fluid column is 2 m. A pressure difference δp raises the fluid surface in one of the arms by 2 cm. If the pressure difference is removed suddenly, what will be the form of the motion of the liquid column?

9.6. When a tympanist strikes his drum the intensity of sound (the loudness) decreases to a half of its initial value of half a second. If the fundamental tone of the drum has a frequency of 200 Hz, make an estimate of the quality factor of the drum.

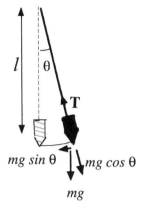

$mg \sin \theta$ $mg \cos \theta$

mg

Fig. 9.7 The forces acting on a plumb line displaced from its equilibrium position through an angle θ.

Solutions

9.1. When the string hangs vertically down, an imaginary extension of its length passes through the centre of mass of the bob. Because the string is specified as light, this is in effect the centre of mass of the plumb line. Take the distance from the point of support to this centre of mass as l. When the sideways displacement occurs the centre of mass will be still at a distance l from the point of support and the bob will move along the arc of a circle of radius l. If the angle of deflection is θ, then the forces acting through the centre of mass are the tension **T** in the string and the weight, mg, of the bob, the mass of which is m. The weight can be resolved into two components, $mg \cos \theta$ acting along the line of the string and $-mg \sin \theta$ perpendicular to the line of the string. The component collinear with the string acts to maintain the tension in the strings while the component perpendicular to the line of the string tends to restore it to its equilibrium from its equilibrium position. For the angular deflection θ, the centre of mass of the plumb line is a distance $s = l\theta$ along the circular arc away from its equilibrium position. The velocity of the centre of mass will be

$$v = \mathrm{d}s/\mathrm{d}t = l(\mathrm{d}\theta/\mathrm{d}t)$$

and its acceleration along the arc will be $\mathrm{d}^2s/\mathrm{d}t^2 = l(\mathrm{d}^2\theta/\mathrm{d}t^2)$. The equation of motion is then

$$m \frac{\mathrm{d}^2 s}{\mathrm{d}t^2} = ml \frac{\mathrm{d}^2 \theta}{\mathrm{d}t^2} = -mg \sin \theta.$$

When θ is small, the approximation $\sin \theta = \theta$ is acceptably accurate (for example, for $\theta = 10°$, θ is different from $\sin \theta$ only by about 0.5%) so that the equation of motion takes the form

$$\frac{\mathrm{d}^2 \theta}{\mathrm{d}t^2} = -\frac{g}{l} \theta = -\omega^2 \theta,$$

which is the equation of motion for a simple harmonic oscillator. The period of oscillation will be $\mathcal{T} = 2\pi/\omega = 2\pi(l/g)^{1/2}$.

This plumb line is a particular example of an oscillating system consisting of a mass carried at the end of a light string (or a light rod), supported from the other end of the string. Such an oscillator is described usually as a simple pendulum.

9.2. The equation of motion for this lossy harmonic oscillator (damped harmonic oscillator) will have the form

$$m\frac{d^2\theta}{dt^2} + b\frac{d\theta}{dt} + \frac{mg}{l}\theta = 0$$

or

$$\frac{d^2\theta}{dt^2} + p\frac{d\theta}{dt} + \omega_0^2\theta = 0,$$

with $p = b/m$ and $\omega_0^2 = g/l = 39.4\,\text{s}^{-1}$. The decay of amplitude is given by

$$A(t) = A(0)[\exp(\tfrac{1}{2}pt)] \qquad \text{or} \qquad \ln[A(0)/A(t)] = \tfrac{1}{2}pt$$

so that, using the figures given,

$$\ln(5/4.5) = \tfrac{1}{2}p \times 1800$$

or

$$p = 1.18 \times 10^{-4}\,\text{s}^{-1}.$$

The time constant for the loss of energy is $\tau = 1/p = 8452\,\text{s}$, around 2 h 22 min. The time constant for the loss of amplitude is $\tau_A = 2/p = \tfrac{1}{2}\tau$, about 4.75 h. Since $p = b/m$, we may evaluate b as $b = 1.18 \times 10^{-3}\,\text{g s}^{-1}$.

The period of the damped oscillation is given by $\mathcal{T} = 2\pi/\omega$ with $\omega = (\omega_0^2 - \tfrac{1}{4}p^2)^{1/2}$. Since $(\tfrac{1}{4}p^2) = 3.5 \times 10^{-9}$ and $\omega_0^2 = 39.5$, we may neglect the effect of damping and say that $\omega = \omega_0$. The period of oscillation is unaffected by the damping.

If, for example, the damping were to have a 1% effect on the period of oscillation, ω would be 1% different from ω_0, so that $\tfrac{1}{2}p$ would be equal to $0.01\,\omega_0$; i.e., $p = 0.79$ and $b = 7.9\,\text{g s}^{-1}$. The time constant for the amplitude would be $\tau_A = 2/p = 2.5\,\text{s}$, so that in 2.5 s the amplitude would have diminished to 1.84°. After 10 s it would be only 0.25°, so that the oscillation would be 'damped out' in only 10 s. There is a natural tendency to think of this oscillation as fairly heavily damped motion, even though the damping is not nearly large enough for the system to be critically damped. The term 'underdamped' is used often to describe systems in which $\omega_0^2 > \tfrac{1}{4}p^2$.

9.3. The static deflection of the springs when the passengers enter the vehicle may be represented by $mg = Kx = 4kx$, since the combined spring constant K is the sum of the individual spring constants, k (springs in parallel). This leads to $k = (320 \times 10)/(4 \times 0.05) = 16\,\text{kN m}^{-1}$. The angular frequencies are obtained from $\omega^2 = k/m$, so that in the unloaded state

$$\omega_u = [(16 \times 10^3)/(10^3)]^{1/2} = 4,$$

and in the loaded state $\omega_1 = [(1.6 \times 10^3)/(1.32 \times 10^3)]^{1/2} = 3.48$. The frequencies are then

$$n_u = \omega_u/2\pi = 0.637\,\mathrm{s}^{-1} \qquad \text{and} \qquad n_1 = \omega/2\pi = 0.554\,\mathrm{s}^{-1}.$$

The damping constant is obtained from the value of the time constant for the amplitude since $\tau = \frac{1}{2}p = 2m/b$, and in one oscillation (in one time period $\mathcal{T}$),

$$A_1/A_2 = \exp(-\mathcal{T}/\tau) = \exp(-\tfrac{1}{2}p\mathcal{T}).$$

The reduction required in the amplitude is 63% in one oscillation, so that

$$A_1/A_2 = 1 - 0.63 = 0.37 = \exp(-\mathcal{T}/\tau),$$

with the consequence that $\tau_u = \mathcal{T}_u = 1.57\,\mathrm{s}$ and $\tau_1 = \mathcal{T}_1 = 1.81\,\mathrm{s}$. The damping constants are then $b_u = 2m/\tau_u = 1273\,\mathrm{kg\,s}^{-1}$ and $b_1 = 1463\,\mathrm{kg\,s}^{-1}$.

The frequencies of the damped oscillations are given by

$$\omega = \left[\omega_0^2 - \left(\tfrac{1}{4}p^2\right)\right]^{1/2} = \left[\omega_0^2 - (1/\tau)^2\right]^{1/2}.$$

In the unloaded state ω is about 4% different from ω_0 and in the loaded state the difference is only about 2.5%.

For critical damping $\omega_0^2 = (1/\tau)^2$, so that in the unloaded case

$$b_u = 2m/\tau_u = 2m\omega_0 = 8000\,\mathrm{kg\,s}^{-1}$$

and in the loaded case $b_1 = 9187\,\mathrm{kg\,s}^{-1}$.

Looking back at the conditions for the overdamped, critically damped, and underdamped systems, we see that these may be expressed as

$$\omega_0 < 1/\tau, \qquad \omega_0 = 1/\tau, \qquad \text{and} \qquad \omega_0 > 1/\tau$$

respectively. These statements could have been written equally as well as $\omega\tau \gtrless 1$, making it possible to describe the system in terms of $\omega_0\tau = Q$, the quality factor of the system. If $Q > 1$ the system exhibits oscillatory behaviour, while if Q is less than or equal to zero the system's behaviour is non-oscillatory. A system with a high value of its quality factor, Q, shows oscillatory behaviour for a long time because dissipative effects are small.

9.4. The amplitude of the pendulum oscillations is reduced from 5° to 1.84° in 100 s. The ratio $1.84/5 = 0.368$ is different from $1/\mathrm{e} = 0.3679$ only by 0.03%. The time constant for the decay of amplitude of the pendulum oscillations, τ, is the time in which the amplitude reduces by a factor $1/\mathrm{e}$, so that we may take $r = 100\,\mathrm{s}$. The ratio of successive maximum amplitudes is then

$$A(1)/A(2) = \exp(\mathcal{T}/\tau) = \exp(0.01) = 1.01$$

or $A(2)/A(1) = 0.99$. The amplitude is reduced by 1% per oscillation so that the energy is reduced by $1 - 0.99^2$ in the same time, giving an energy loss of 2% per oscillation. To maintain the oscillations at an amplitude of 5°, energy has to be supplied at a rate of

$$\mathrm{d}E/\mathrm{d}t = 0.02\left\{\tfrac{1}{2}m[A(1)]^2\,\omega^2\right\}$$

per oscillation, in this case per second. $\omega = 2\pi/\mathcal{T}$, so that $\omega^2 = 4\pi^2 = 39.45$. The amplitude, $A(1)$, of the oscillation is the product of the angular amplitude, θ_0, and the length, l, of the pendulum. The equivalent length of the pendulum is obtained from the simple pendulum relation $\mathcal{T} = 2\pi(l/g)^{1/2}$, and has the value 24.85 cm. Then $A(1) = l\theta_0 = 0.2485(8.73 \times 10^{-3}) = 2.17 \times 10^{-2}$ m. Finally,

$$dE/dt = 0.02(\tfrac{1}{2} \times 0.02 \times 4.7 \times 10^{-4} \times 39.48) = 3.71 \times 10^{-5} \,\text{J s}^{-1}.$$

The clock weight has a velocity, $v = (1.5)/(7 \times 24 \times 3600) = 2.48 \times 10^{-6}\,\text{m s}^{-1}$, and produces energy at a rate of $mgv = 6.2 \times 10^{-5}\,\text{J s}^{-1}$. The proportion of that energy required to maintain the amplitude of oscillation is $3.71/6.2 = 0.598$, practically 60%.

9.5.   Let's say that the pressure difference p raises the fluid surface from its initial position, h, by a distance y up one arm. Because the quantity of fluid is fixed, the fluid surface in the other arm is lowered by the same distance. The pressure difference maintains a disparity, δm, between the masses of fluid in the two arms. In this equilibrium state the force exerted by the pressure difference is nullified by the weight $g\delta m$ so that $g\delta m = pA$, where A is the area of the tube internal cross-section. δm is the product of the density of the fluid and the difference between the volumes of fluid contained in the two arms, i.e. $\delta m = 2\rho g A y$, with the result that $p = 2\rho g y$. When the pressure difference is changed suddenly to zero, the force acting on the fluid column is simply $g\delta m = 2\rho g y A$, and this force is exerted on the fluid column, the mass of which is $\rho A l$, where l is the total length of the fluid column. The equation of motion is then

$$\rho A l \frac{d^2 y}{dt^2} = -2\rho g A y$$

or

$$d^2 y/dt^2 = -2(g/l)y,$$

which is the equation of a simple harmonic motion of angular frequency $\omega = (l/2g)^{1/2}$, and which—interestingly—is independent of the initial difference in height. Here $l = 2$ and $\omega = 0.32\,\text{s}^{-1}$.

Such a manometer wouldn't be of much use, because without dissipative forces the harmonic motion would continue for ever. Fortunately, almost all fluids show a resistance to flow, even when that flow is not turbulent, and this is a manifestation of their viscosity. The viscosity of the fluid in the manometer will provide a dissipative mechanism by means of which the oscillations of the fluid column will be damped. The 'best case scenario' would be when the system was critically damped so that the relaxation time for the amplitude of fluid oscillations would be $\tau = 1/\omega_0$. In the present case the value of τ for critical damping would be 3.1 s. This observation provides one of the reasons why the choice of fluid in a manometer is not simply arbitrary. Overdamping would lead to a longer relaxation time, while underdamping would lead to undesirable oscillatory behaviour.

9.6. The sound waves which are generated by the drum are the result of a transfer of energy from the vibrating surface of the drum to the surrounding air. If we model the drum as a linear oscillator, we may deal with the effect of sound wave generation by introducing a dissipative term into the equation of motion of the oscillator. Then the rate of energy loss by the drum is given by

$$dE/dt = -\tfrac{1}{2}Ep = -E/\tau,$$

so that

$$E(t) = E(0)\exp(-t/\tau).$$

The intensity of a wave is the rate at which energy is transferred by the wave, and for an observer at a fixed distance from the drum this will be directly proportional to the rate at which the energy of the drum is dissipated. This means that the intensity of the sound heard by the observer is $I(t) = I(0)\exp(-t/\tau)$, so that in this case $0.5 = \exp(-t/\tau)$, or $\tau = 0.72$ s. The fundamental frequency of the drum is 200 Hz, which is an angular frequency of $1257\,s^{-1}$, so that $Q = \omega_0\tau = 906$. This large value of the quality factor is directly proportional to the frequency of the sound. If that frequency were reduced to 50 Hz and the intensity of sound decreased at the same rate the value of Q would be 226.5. Describing the drum in terms of its fundamental tone is a simplification which is hardly justifiable when it is remembered that the quality of sound emitted by any musical instrument is determined to a large extent by oscillations of higher frequency than the fundamental, the overtones.

This example acts as a reminder that the transfer of energy from the drum is predominately by means of generation of sound. If the drum were placed in an evacuated space so that the production of sound was no longer a possible dissipative mechanism, would the drum continue to vibrate for ever? The answer has to be negative. First, there is the possibility that the energy of the drum would be transferred to the walls of the enclosure in which it is placed. If it stood on the floor of the vacuum chamber energy could be transferred through the legs of the drum to the floor. Even if the drum was suspended by fine strings so that it was effectively isolated from the enclosure (the case of an adiabatic enclosure) the vibrations would die away eventually. Energy would be dissipated by 'internal friction' in the drum skin and sides. The internal friction mechanism would transform the vibrational energy to thermal energy. Since the effects of internal friction are generally very small, the quality factor would be expected to have a very high value when the drum was in an adiabatic enclosure.

10
Rotational motion—torque

10.1 Motion about the centre of mass

In effect, we have dealt so far only with the motion of the centres of mass of bodies. We have treated the systems under consideration simply as if all their mass has been concentrated at their centres of mass. This has allowed us to think of them as massy point particles that can't have any motion with respect to their centres of mass. An everyday example showing that this approach is not always adequate is that of a bicycle wheel. If the wheel is lifted off the ground it is free to rotate about its axle and does so when subjected to a suitable impulse. The axle, which is effectively the centre of mass of the wheel, stays in the same place, so that the motion of the wheel consists only of motion with respect to the centre of mass. There is no kinetic energy of translation associated with movement of the centre of mass, but there is energy associated with the wheel's rotation about the axle. On the other hand, when the wheel rotates while it is in contact with the ground, it progresses along. The frictional force between the ground and the wheel moves the wheel along its path. Its centre of mass (at the axle) moves so that there is translational motion of the centre of mass as well as the rotational motion relative to the centre of mass. There is energy of translational as well as of rotational motion. If the bicycle were pushed along its path it could be said that the frictional force between wheel and ground causes the rotation of the wheel, leading us to ask about the relationship between force and motion with respect to the centre of mass. Obviously, if the centre of mass is static, it's impossible to use the identity of force and the product of mass and acceleration because the acceleration of the centre of mass is zero. To begin, we will deal with situations in which the centre of mass of the body under consideration does not move.

Example 56. A light rod rests on a pivot passing through its centre of mass. A force F_1 acts vertically down on the rod to the right of the pivot point at a distance x_1 from the pivot. A force F_2 acts vertically downwards on the rod at a distance x_2 to the left of the pivot. What is the condition, in terms of the F's and x's, that the rod does not rotate about its pivot point?

The situation of the rod is depicted in (Fig 10.1(a)). Forces F_1 and F_2 act vertically downwards, and there must be a force F acting vertically upwards equal in magnitude to their sum; otherwise, the centre of mass would move. There will be a tendency for the rod to rotate about O unless F passes through O. The resolution of this problem is simplified if we introduce a fictitious force f at either end of the rod, acting along the length of the rod. Whatever the magnitude of f, the force that results from adding together the f's and F's must pass through a point P vertically above O; otherwise, the rod will rotate. Let the lines of action of

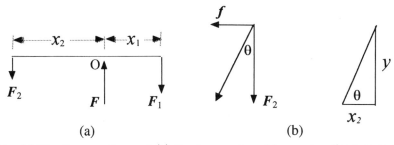

Fig. 10.1 (a) The forces acting and (b) the force and position vector diagrams to find the 'no rotation' condition.

the combination of f and F_1 and of the combination of f and F_2 meet at a height y above O. The force vector diagram for the left-hand forces and the position vector diagram for the action of those forces are given in Fig. 10.1(b). From the force diagram, $|f|\tan\theta = |F_2|$ and from the position diagram $\tan\theta = (y/x_2)$. Then $|f| = |F_2|(x_2/y)$. Similarly, for the right-hand side forces, $|f| = |F_1|(x_1/y)$. On equating these, we see that P is vertically above O, i.e. the resultant force passes through O, provided that

$$F_1 x_1 = F_2 x_2,$$

where $F_i = |F_i|$.

The product of the force and its distance from O, the axis about which rotation could take place, is called the moment of the force about O, or the couple resulting from the force, or the torque, Γ, produced by the force. Obviously, from Fig 10.1, $\Gamma_1 = F_1 x_1$ acts to cause rotation in a clockwise direction and Γ_2 acts in an anticlockwise direction. Had we taken O as the origin of a set of Cartesian axes, x_2 would have been negative, showing the opposite senses of the two torques. The condition of equilibrium would then have been $\Gamma_1 + \Gamma_2 = 0$ or, more generally, $\Sigma\Gamma_i = 0$. This general statement gives us a second condition which must be satisfied for a system to remain static; i.e. for a system to be in static equilibrium, $\Sigma\Gamma_i = 0$ and $\Sigma F_i = 0$. A system will be in static equilibrium only if the nett force acting on it is zero and the nett torque acting on it is zero.

10.2 Equations of angular motion

Example 57. The two-bladed propellor on an aircraft may be modelled (very crudely) as a uniform rod of length l and mass M with a shaft passing through its centre around which it can rotate. If the propellor is set in uniform rotation with an angular velocity ω, what is the kinetic energy associated with the rotary motion?

The rotating propellor is shown schematically in Fig 10.2. The kinetic energy of an element of mass δm situated at P, distant r from the shaft, is simply $\frac{1}{2}v^2\,\delta m$. v is the tangential velocity of δm as it rotates around its circular path and $v = r\omega$, so that the contribution of the mass element to the kinetic energy of the propellor is

$$\delta T = \tfrac{1}{2}r^2\omega^2\,\delta m.$$

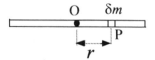

Fig. 10.2 The rotation of a rod about a shaft passing through its centre.

The total kinetic energy of the propellor is obtained by summing T over all the values of r, i.e.,

$$T = \sum \delta T = \lim_{\delta m \to 0} \sum \tfrac{1}{2} r^2 \omega^2 \, \delta m$$

$$= \tfrac{1}{2} \omega^2 \int r^2 \, dm,$$

where the summation is expressed as an integral because $r^2 \delta m$ is a continuously varying quantity. We can be more specific than this and evaluate the integral. If the element of mass δm extends a distance δr in the r direction, then $m = (M/l)r$ and so, for $-\tfrac{1}{2}l \leqslant r \leqslant \tfrac{1}{2}l$,

$$T = \tfrac{1}{2} \omega^2 (M/l) \int r^2 \, dr = \tfrac{1}{2}(M/l)\omega^2[l^3/12]$$

$$= \tfrac{1}{2} \omega^2 [(Ml^2)/12].$$

The quantity $(Ml^2)/12$, derived during the integration, is characteristic of the rod and the position of the shaft about which it rotates. If the shaft were situated at one end of the rod, then the integral would have the limits of 0 and l, and the new form of multiplier for $\tfrac{1}{2}\omega^2$ in the expression for the kinetic energy of rotation would be $(Ml^2)/3$.

Say that one end of the propellor is subjected to a tangential force F. When the shaft is at the centre of the propellor this produces a torque of $\tfrac{1}{2}Fl$. What is the equivalent force, $F(r)$, acting on the element of mass δm? It's simply the force that produces the same torque, i.e. $rF(r) = \tfrac{1}{2}Fl = \Gamma$, or $F(r) = F(l/2r)$. If we assumed that the element were free to move on its own, we could say that this tangential force would produce a tangential acceleration (dv/dt) such that

$$F(r) = Fl/2r = \delta m(dv/dt).$$

Since $v = r\omega$, this may be written as

$$\Gamma = \tfrac{1}{2}Fl = r^2 \, \delta m(d\omega/dt).$$

However, the element is not free to move on its own: it has to move in conjunction with the rest of the propellor. All of the mass elements which constitute the propellor move together. Because of this, the angular acceleration that the propellor will suffer has to take account of the summation of all the $r^2 \delta m$ as $\delta m \to 0$ and

$$\Gamma = (d\omega/dt) \int r^2 \, dm.$$

Since r is continuous between the limits $\frac{1}{2}l$ and $-\frac{1}{2}l$,

$$\Gamma = (d\omega/dt)[Ml^2/12].$$

What is obvious from this equation of motion for rotary motion is that the quantity $\int r^2\, dm$ has a characteristic form which depends on the axis of rotation and on the shape of the rotating body. It is called the moment of inertia, I, and

$$I = \lim_{\delta m \to 0} \sum r^2\delta m = \int r^2\, dm$$

is the moment of inertia of the body about the particular axis being considered. It could be called the second moment of the mass of the body. The moment of inertia of a body can also be defined as the product of the mass of the body and the square of its radius of gyration, K, i.e. $I = MK^2$. The equation of motion can be generalized as

$$\Gamma = I(d\omega/dt) = I\omega(d\omega/d\theta) = I(d^2\theta/dt^2)$$

since, with angular displacement θ, $\omega = (d\theta/dt)$ and $(d\omega/dt) = (d\omega/d\theta)(d\theta/dt)$. The similarities between these equations and those that we established for linear motion suggest that we could define an angular momentum $L = I\omega$, so that

$$\Gamma = dL/dt.$$

One intriguing point about this definition of angular momentum is that we have a particle of mass m moving with angular velocity along a circular arc of radius r, then its angular momentum is $mr^2\omega = mvr$, in terms of the tangential velocity. The angular momentum is not simply the tangential linear momentum, mv, expressed in terms of r and ω. It is the moment of that momentum about the centre of rotation. For this reason, angular momentum is often called the moment of momentum.

The definition of angular momentum leads us to the concept of angular impulse, the change in angular momentum δL brought about by the application of a torque Γ for a time interval dt

$$J_\omega = \int \Gamma\, dt = \delta L.$$

The analogy between the quantities defined for linear motion and these for angular motion can be used to define the work done by a torque rotating through an angle θ, as

$$W = \Gamma\theta$$

and the power exerted by a torque (the rate of working), as

$$P = \Gamma\omega.$$

10.3 Moments of inertia

Example 58. What is the moment of inertia of a circular disc of radius a about an axis perpendicular to its plane, passing through its centre?

This is a particularly simple example, because of the symmetry of the disc about the chosen axis. The mass element is made up from elements of thickness dr at a distance r from the axis, O, as shown in Fig 10.3. The element has the form of an annulus of radius r and thickness δr so that $\delta m = \sigma\, 2\pi r\, dr$, where σ is the mass per unit area of the disc. Then

$$I = \sum r^2\, \delta m$$

which as $\delta m \rightarrow 0$ takes the form

$$I = 2\pi\sigma \int_0^a r^3\, dr = 2\pi\sigma \tfrac{1}{4}a^4$$

The area of the disc is a^2, so that its mass is $M = \pi a^2 \sigma$ and

$$I = \tfrac{1}{2}Ma^2.$$

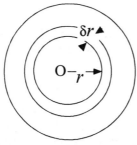

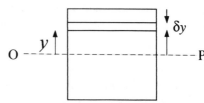

Fig. 10.3   An illustration of the calculation of the moment of inertia of a circular disc.

Fig. 10.4   Illustrating the calculation of the moment of inertia of a square lamina about the axis OP.

Example 59.   A uniform square lamina has sides of length l. What is its moment of inertia about an axis in its plane which passes through the centres of two opposite sides?

Once again, the axis chosen matches well with the symmetry of the lamina. We can choose rectangular elements of mass distant y from the axis, and of thickness δy, as shown in Fig 10.4. If the lamina has mass M, its mass per unit area $\sigma = M/l^2$, and then

$$\int r^2\, dm = \sigma l \int y^2\, dy$$

between the limits $\tfrac{1}{2}l$, giving

$$I = \sigma l(l^3/12) = M(l^2/12).$$

Not surprisingly, this is the same as the moment of inertia of the rod in Example 57. Had we looked at the lamina along the direction of the axis OP, we would have seen a rod with an axis passing through its centre. The extent of the lamina in the direction parallel to the axis OP is not relevant to our result, except that it defines the mass per unit area.

Example 60. A simple pendulum consists of a point mass m suspended at the end of a light rod of length l. Using the equations for angular motion, show that the period of oscillation, $\mathcal{T}$, describing small amplitude oscillations of the pendulum, has the form $2\pi(l/g)^{1/2}$, where g is the acceleration due to gravity.

The light rod can be taken as having no mass, so that the moment of inertia of the rod–mass system about the point of suspension is ml^2. As we have seen in Exercise 9.1, the tangential force which tends to return the pendulum bob to its original position is $-mg\sin\theta$, so that the torque acting on the system is $-mgl\sin\theta$ which, for small values of θ, can be identified with $-mgl\theta$. Equating the torque, $\Gamma = -mgl\theta$, to the product of moment of inertia and angular acceleration gives

$$\Gamma = I\frac{d^2\theta}{dt^2} = -mgl\theta = -ml^2\frac{d^2\theta}{dt^2},$$

which is the equation of motion of a linear oscillator the period of oscillation of which is given by $\mathcal{T} = 2\pi(l/g)^{1/2}$.

10.3.1 The parallel axis theorem

The period of oscillation could have been written as $2\pi(I/mgl)^{1/2}$, which is a generalization of the result obtained for the simple pendulum. This moment of inertia is taken about the point of suspension, not about the centre of mass as in Examples 58 and 59. If the point of suspension coincided with the centre of mass there would be no oscillatory motion when the body was displaced: it would rotate about the centre of mass or (if there were a dissipative torque) move to a new position of equilibrium. We have seen that it is convenient, for the calculation of moments of inertia, for the axis about which the calculation is made to coincide with the centre of mass, and are led to ask about the relationship between the moment of inertia about some axis passing through the centre of mass of a body and another axis, parallel to the first one but situated away from the centre of mass. This situation is illustrated in Fig 10.5, in which the axes are taken as perpendicular to the plane of the drawing and pass through the centre of mass C and the arbitrary position A. A mass element δm situated at a point B in the

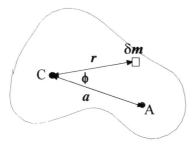

Fig. 10.5 A diagram showing the contribution of a mass element δm to the moments of inertia about axes perpendicular to the plane of the diagram, passing through the centre of mass, C, and through the point A.

body, at a distance r from C, contributes $r^2 \delta m$ to the moment of inertia about C, and $y^2 \delta m$ to the moment of inertia about A. The cosine rule enables us to relate these, by

$$y^2 = r^2 + a^2 - 2ar \cos \phi,$$

so that

$$y^2 \, \delta m = r^2 \, \delta m + a^2 \, \delta m - 2ar \, \delta m \cos \phi.$$

In Figure 10.5 we may identify CA as the x-axis, with C as origin, giving $x = r \cos \phi$. Then

$$\sum ar \, \delta m \cos \phi = \sum ax \, \delta m$$

which in the limit as $\delta m \to 0$ becomes $\int ax \, dx = aMX$ where M is the mass of the body and X the distance of its centre of mass from C. Since C is the centre of mass, $X = 0$ and

$$\sum \delta m y^2 = \sum \delta m r^2 + \sum \delta m a^2$$

or

$$I_A = I_C + Ma^2$$
$$= M(K_C^2 + a^2),$$

where K_C is the radius of gyration about the centre of mass. This theorem of parallel axes has been illustrated implicitly in Example 57, in which the moments of inertia of a uniform rod about its centre of mass and about one of its ends were worked out. Example 60 could be said to use the theorem as well, although in a very elementary way.

10.3.2   The perpendicular axis theorem

There is a second method of simplifying calculations of moments of inertia which can be applied in the case of laminar bodies. It is known as the theorem of perpendicular axes. This is illustrated in Fig 10.6, in which is shown a set of Cartesian axes, two of which—say the x- and y-axes—lie in the plane of the lamina. The mass element δm at the point P contributes $x^2 \, \delta m$ to the moment of

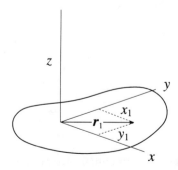

Fig. 10.6   The axes and their position with respect to the lamina in the perpendicular axis situation.

inertia about the y-axis, $y^2 \, \delta m$ to the moment of inertia about the x-axis, and $r^2 \, \delta m$ to the moment of inertia about the z-axis which is perpendicular to the plane of the lamina. Obviously,

$$r_1^2 = x_1^2 + y_1^2,$$

so that the moments of inertia about the z, y-, and x-axes are related by

$$\sum \delta m r^2 = \sum \delta m x^2 + \sum \delta m y^2$$

or

$$I_Z = I_Y + I_X.$$

Example 61. A shop sign is in the form of a circular disc of radius a, suspended by a horizontal hinge which is tangential to the disc. What would be the period of oscillation of the sign if it suffered a small displacement from its equilibrium position?

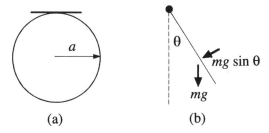

(a)   (b)

Fig. 10.7 (a) The circular shop sign. (b) The force acting on the sign when it is displaced through a small angle.

The shop sign is shown in Fig 10.7(a). When it suffers its displacement it is subjected to a restoring torque due to its weight, as shown in Fig 10.7(b). As in Example 60, the torque acting to return it to its equilibrium position is $Mga \sin \theta$, or for small angles simply $Mga\theta$. The equation of motion is then

$$\Gamma = -Mga\theta = I \frac{d\omega}{dt},$$

so that the solution of the problem devolves into the evaluation of the moment of inertia of the disc about a tangent lying in its plane. This can be resolved quite simply because of the symmetry of the disc and the two theorems given in the two preceding sections. In Example 58 we found that the moment of inertia of a disc about an axis perpendicular to its plane and passing through its centre of mass was $\frac{1}{2}Ma^2$. The symmetry of the disc tells us that the moment of inertia of the disc about a diameter is the same irrespective of the orientation of the diameter in the plane of the disc. The moments of inertia, I_X and I_Y, about two perpendicular diameters will be identical so that the moment of inertia, I_Z, of the disc about an axis passing through its centre of mass perpendicular to its plane will be

$$I_Z = I_X + I_Y = 2I_X,$$

so that

$$I_X = I_Y = \tfrac{1}{2}I_Z = \tfrac{1}{4}Ma^2.$$

The moment of inertia about a tangent, distant a from a diameter, will be

$$I = I_X + Ma^2 = \tfrac{5}{4}Ma^2.$$

The period of oscillation

$$\mathcal{T} = 2\pi(I/Mga)^{1/2} = 2\pi(5a/4g)^{1/2}.$$

This example could have been solved just as well by calculating the moment of inertia about a diameter directly and then using the parallel axis theorem. The integration would be more complicated, though not difficult, and the method would be analytical rather than (almost) geometrical.

Exercises

10.1.   A flywheel in the form of a heavy circular disc of mass 10 kg with a diameter of 30 cm is mounted on a light axle passing through its centre of mass, so that it can rotate freely in a vertical plane. If a mass of 5 kg is attached to the highest point of the disc and the disc–mass system is displaced slightly from its position of (unstable) equilibrium, what is the angular velocity of the flywheel when the 5 kg mass is at its lowest point?

10.2.   Miners are lowered to the levels at which they work in a pit cage. This is a rudimentary elevator, in which the suspension carrying the cage is wound round a drum of 3 m diameter and 1 tonne mass. To start the cage on its descent the winder simply releases the brake on the drum, which then rotates freely. If the cage and its contents have a mass of 1 tonne, what are the initial acceleration of the cage, the initial angular acceleration of the drum, and the tension in the cable?

10.3.   A well windlass has a spindle of radius 10 cm and a handle which turns in a circle of 50 cm radius. It is used to raise a 20 kg bucket full of water through a height of 10 m. What is the average force used to raise the water?

10.4.   A bicycle of mass 15 kg is travelling at a speed of 36 kph. Its wheels have a radius of 24 cm and their masses are 1.5 kg each, effectively all at the rim of the wheels. What is the kinetic energy of the bicycle? If it has travelled 50 m from rest to attain its speed, what is the average force which has been used to accelerate it to 36 kph?

10.5.   Three small boys have a marble, a golf ball, and a table tennis ball amongst them. Their competitive spirit leads them to hold a rolling race in which these three spherical objects are released simultaneously from the top point of a baize-covered sloping plane (eg. a miniature pool table) tilted at an angle of 30° to the horizontal, and allowed to roll freely down the slope. Which of these objects will be first to reach the bottom of the slope? If the slope is 60 cm in length, what time intervals will separate the arrivals of the three objects?

10.6.   A toy car running at a speed of $1\,\text{m s}^{-1}$ has its power supplied by a flywheel which has a mass of 100 g and can be regarded as a uniform circular disc of radius 3 cm rotating initially at 1200 rpm. If the toy requires a power of 22.2 mW to maintain its speed at $1\,\mathrm{ms^{-1}}$, what will be the rate of rotation (in rpm) of the flywheel when the toy has travelled a distance of 3 m?

10.7. A boy, walking across the surface of a frozen pond, entertains himself by kicking a straight uniform stick, 1 m long and of mass 1 kg, which lies on the ice. If his kick provides an impulse of 2 Ns at one end of the stick, what is the form of the subsequent motion of the stick?

10.8. A girl throws a snowball at a pub sign which is in the form of a rectangle of height 1.5 m, width 1 m, and of mass 12 kg, hanging vertically from a hinge along its top edge. Her aim is good and the snowball, of mass 100 g, is travelling with a horizontal component of velocity of 80 kph when it strikes the sign 20 cm above the centre of its bottom edge. What is the impulse which acts at the hinge when the snowball strikes the sign? What will be the angular velocity of the sign immediately after the impact?

10.9. In a bowling alley, a bowler projects a bowling ball, of mass 8 kg and diameter 20 cm, with a velocity of $9\,\mathrm{ms^{-1}}$ along the bowling lane towards the skittle pins which are 18 m from the bowler. Make an estimate of the velocity of the ball when it arrives at the pins if the coefficient of kinetic friction between the ball and the floor is 0.2. What difference would it make if the bowler were able to impart spin about its horizontal axis to the ball as he projected it?

Solutions

10.1. During its descent from its highest to its lowest position, the 5 kg mass suffers a vertical fall of one disc diameter, so that it loses potential energy $mgh = 0.3 \times 5g$ J. This potential energy has been transformed to kinetic energy, $\frac{1}{2}mv^2$, of the mass and kinetic energy of rotation, $\frac{1}{2}I\omega^2$, of the disc, where v is a translational velocity and ω an angular velocity. Since if a is the disc radius, $I = \frac{1}{2}Ma^2 = \frac{1}{2} \times 10 \times (0.15)^2$ and $v = 0.15\omega$, on equating potential energy loss to kinetic energy gain

$$0.15g = \tfrac{1}{2}[5 \times (0.15)^2]\omega^2 + \tfrac{1}{4}[10 \times (0.15)^2]\omega^2$$

leading to $\omega = 8.08\,\mathrm{s^{-1}}$ ($= 8.08\,\mathrm{radians\,s^{-1}}$).

Instead of distinguishing between the rotational and translational kinetic energies of the disc and the mass, we could have said that the kinetic energy of rotation was the sum of the rotational kinetic energies of the disc and the mass. The moment of inertia of the mass was simply $ma^2 = 5 \times (0.15)^2$, so that its contribution to the rotational energy is $\frac{1}{2}[5 \times (0.15)^2]$ J, just the same as in the energy equation above. This observation highlights the statement that if two bodies are joined rigidly together and their combination can rotate about an axis then the moment of inertia of the combination about that axis is the sum of their individual moments of inertia about that axis.

This is self-evident from the definition of moment of inertia, $I = \sum \delta m r^2$, but bears repeating.

10.2. The elevator is shown schematically in Fig 10.8, together with the forces, the weight, mg, of the cage and the tension, T, in the cable. The net force acting on the cage is

$$mg - T = m(dv/dt)$$

and the torque acting on the drum is

$$Tr = I(d\omega/dt) = \tfrac{1}{2}Mr^2(d\omega/dt),$$

where r is the drum radius and M is its mass. We may divide this torque equation by r, and identify $v = r\omega$, giving

$$T = \tfrac{1}{2}Mr(d\omega/dt) = \tfrac{1}{2}M(dv/dt).$$

Then

$$mg - \tfrac{1}{2}M(dv/dt) = m(dv/dt)$$

or

$$\frac{dv}{dt} = \frac{mg}{(m + \tfrac{1}{2}M)},$$

so that the initial acceleration of the cage is $\tfrac{2}{3}g$, about $6.5\,\text{m}\,\text{s}^2$. The angular acceleration of the winch is $\tfrac{2}{3}g/r = 4g/9$, about $4.4\,\text{s}^{-2}$. The tension in the cable is $3.2\,\text{kN}$.

The acceleration of the cage is less than g because the weight of the cage does work in accelerating the drum as the cable unwinds from it. We have neglected the mass of the suspension cable and the variations in diameter and moment of inertia of the drum as the cable unwinds from it. This is justifiable in making an estimate of the initial accelerations and tension.

10.3. The bucket lifting mechanism is sketched in Fig 10.9, together with the

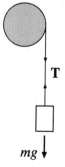

T

mg

Fig. 10.8 A schematic diagram of the pit cage/winding assembly, showing the forces of tension, T, and of the weight, mg, acting on it.

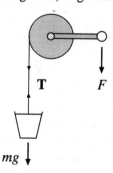

T F

mg

Fig. 10.9 A sketch of the bucket lifting mechanism of a well. The forces acting in the system are the weight of the bucket, mg, the tension in the rope T, and the force, F, applied to the handle.

forces acting on it. When the bucket is raised its potential energy increases by $mgh = 20 \times 10 \times 9.8 = 1960$ J. During the lift a 10 m length of rope has been wrapped around the spindle of radius 0.1 m, so that the number of revolutions made by the spindle is

$$n = 10/0.2\pi.$$

The spindle has turned through $2\pi n$ radians, i.e. the angle of turn is 100 radians. The handle has turned through the same angle so that the work done by the torque, Fr ($= 0.5F$), acting on it is $0.5F \times 100 = 50F$. This work must be equal to the change in potential energy of the bucket so that $50F = 1960$, or $F = 39.2$ N. The ratio of the radius of the circle swept out by the handle to the radius of the spindle (which is the same as the ratio mg to F) is known as the mechanical advantage of the windlass.

10.4. The kinetic energy of the bicycle is a combination of its translational kinetic energy and the rotational kinetic energy of its wheels, i.e.

$$T = \tfrac{1}{2}Mv^2 + 2\left(\tfrac{1}{2}I\omega^2\right),$$

where v is the velocity of the bicycle, M is its mass, I is the moment of inertia of a wheel, and ω is the angular velocity of the wheels. Making the (crude) approximation that the mass of the wheels is concentrated at their rims, $I = mr^2$, and $I\omega^2 = mr^2\omega^2 = mv^2$. The kinetic energy has the value

$$T = \tfrac{1}{2}[15 \times 10^2] + [1.5 \times 10^2] = 900 \text{ J}.$$

The average force exerted to produce the velocity of 36 kph in a distance x is

$$F = T/x = 900/50 = 18 \text{ N}.$$

This is a larger force than would be needed to accelerate the bicycle to the same velocity over a smooth surface (which would be only 15 N). The reason for this lies in the equation

$$Fx = \tfrac{1}{2}[M + (2I/r^2)]v^2,$$

which could be written as $Fx = \tfrac{1}{2}\mu v^2$, where, as far as kinetic energy is concerned, $\mu = [M + (2I/r^2)]$ is the effective mass of the bicycle, here some 20% larger than M.

10.5. The torque producing the rolling motion of the spherical object shown in Fig 10.10 is the product of the contact force, f, and the radius, r, of the sphere,

$$rf = I(\mathrm{d}\omega/\mathrm{d}t),$$

where $\mathrm{d}\omega/\mathrm{d}t$ is the angular acceleration of the sphere. The equation may be rewritten as

$$f = (I/r)(\mathrm{d}\omega/\mathrm{d}t) = (I/r^2)(\mathrm{d}v/\mathrm{d}t),$$

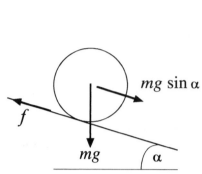

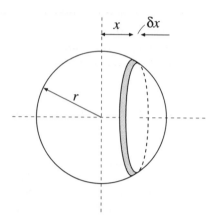

Fig. 10.10 A spherical object rolling down a slope. The forces acting parallel to the slope are the contact force, f, and the component of the weight, $mg \sin \alpha$.

Fig. 10.11 Disc elements which are used to find the moment of inertia of a solid sphere about a diameter.

where v is the linear velocity of the sphere. The equation for the linear motion of the sphere is then

$$m(dv/dt) = (mg \sin \alpha) - f$$

or

$$[m + (I/r^2)](dv/dt) = mg \sin \alpha,$$

so that

$$\frac{dv}{dt} = \frac{1}{[1 + (I/mr^2)]} g \sin \alpha.$$

To find out which of the balls will have the largest acceleration, we need to know their moments of inertia. The marble and the golf ball can be taken (approximately) as solid homogeneous spheres, and the table tennis ball as a hollow spherical shell. The moment of inertia of a solid sphere about a diameter is found by taking disc-shaped mass elements, as shown in Fig 10.11. The mass of a disc element is $\pi \rho y^2 \, \delta x$ where ρ is its density, and its moment of inertia about the x-axis (a diameter) is $\frac{1}{2}\pi \rho y^4 \delta x$. For the whole sphere, $-r < x < r$, and remembering that $x^2 + y^2 = r^2$,

$$I = \tfrac{1}{2}\pi \rho \int y^4 \, dx$$

$$= \tfrac{1}{2}\pi \rho \int (r^2 - x^2)^2 \, dx$$

$$= \tfrac{1}{2}\pi \rho (16 r^5 / 15)$$

$$= (2 M r^2 / 5).$$

The calculation for the hollow sphere is made by using hoop, rather than disc, mass elements. The mass of a hoop element is expressed in terms of an angular element $\delta\theta$ as $2\pi y \times mr\,\delta\theta$ where m is the mass per unit area of the shell. The moment of inertia of the element about the x-axis takes the form

$$\delta I = 2\pi rmy \times y^2\,\mathrm{d}\theta$$

Then

$$I = 2\pi m \int y^3\,\mathrm{d}\theta$$

which, on substituting $y = r\sin\theta$, gives, between the limits of 0 and π,

$$I = 2\pi mr^4 \int \sin^3\theta\,\mathrm{d}\theta$$

$$= (8\pi mr^4/3)$$

which, with $M = 4\pi r^2 m$, become $I = \frac{2}{3}Mr^2$ for a hollow sphere.

On substituting these values of I into the equation for the acceleration, we find that the accelerations of the marble and the golf ball would be

$$a_s = \tfrac{5}{7}g\sin\alpha = 0.71g\sin\alpha = 0.36g.$$

for $\alpha = 30°$, while for the table tennis ball

$$a_h = \tfrac{3}{5}g\sin\alpha = 0.6g\sin\alpha = 0.3g.$$

The golf ball and the marble will accelerate at the same rate and travel a distance of 60 cm in 0.59 s whilst the table tennis ball will arrive at the 60 cm mark about 0.05 s later. Had the slope been smooth so that the balls slipped rather than rolled, then they would all have accelerated at the same rate and would have reached the 60 cm mark after travelling for 0.5 s.

The translational acceleration and any velocity or position which can be deduced refer, of course, to the centre of mass. The times given above are the times for the centres of mass of the spherical objects to travel through 60 cm. As an alternative, it could be said that the points of contact of the objects with the slope travel through that distance in those times. The solidity of the balls would make it difficult to judge the positions of their centres of mass, and the result of the boys' race would depend on their starting procedure and their definition of crossing the 60 cm mark. The golf ball, being of larger diameter than the marble, might appear to arrive first. The boys might (incorrectly) associate the fastest descent with the greatest weight. If we take the point of contact between the surface and the ball as our reference point, the equation of motion takes the form

$$I\frac{\mathrm{d}\omega}{\mathrm{d}t} = r \times Mg\sin\alpha$$

and, since $v = r\omega$,

$$\frac{I}{Mr^2}\frac{\mathrm{d}v}{\mathrm{d}t} = g\sin\alpha.$$

When we recollect that the moment of inertia of the (solid) sphere relative to our point of reference is $\frac{7}{5}mr^2$, we see that the acceleration of the sphere is $\frac{5}{7}g\sin\alpha$, just as before.

The choice of reference point is thus irrelevant to the result. It may affect the ease or difficulty of solving the equations of motion, but the final outcome should be the same independent of the position of the reference point. Here it has not mattered whether we have considered the sphere to rotate with angular velocity ω about its centre of mass or have taken the centre of mass of the sphere as rotating with angular velocity ω about the point of contact of the sphere on the slope.

10.6. Taking the flywheel as a uniform circular disc rotating about an axis passing through its centre, perpendicular to its plane, we can say that its moment of inertia is $I = \frac{1}{2}ma^2 = \frac{1}{2}[0.1 \times (0.015)^2] = 1.125 \times 10^{-5}\,\text{kg m}^2$. The initial energy of the flywheel is $T_1 = \frac{1}{2}I\omega^2$ with $\omega = \text{rpm} \times (2\pi/60)$, i.e. $T = 88.8\,\text{mJ}$. The toy travels 3 m in 3 s, so that the energy expended ($=$ power $\times$ time) is 66.6 mJ.This is provided at the expense of the rotational kinetic energy of the flywheel, which is reduced from 88.8 mJ to $T_F = 22.2\,\text{mJ}$. The angular velocity of the flywheel has been reduced to

$$\omega_F = (2T_F/I)^{1/2} = 62.8\,\text{s}^{-1} = 600\,\text{rpm}.$$

To all intents, the rate of rotation of the flywheel has been halved.

The total energy of the toy is a combination of its translational energy and the rotational energy of the flywheel. Since the initial and final velocities of the toy are the same, we can omit the translational energy from our consideration.

10.7. Let's say that the stick has mass m and length $2a$, as shown in Fig 10.12. Then, if $I(=\frac{1}{3}ma^2)$ is the moment of inertia about the centre of mass of the stick, the impulsive angular torque (the angular impulse) produced by the kick is equal to the change in angular momentum which is observed, i.e.

$$Ja = I\omega$$

or

$$\omega = Jl/a = 3J/ma = 12\,\text{s}^{-1}.$$

The impulse also produces a centre of mass velocity, v, since $J = mv$, or

$$v = J/m = 2\,\text{m s}^{-1}$$

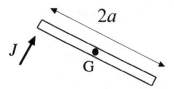

Fig. 10.12 The stick of mass m and length $2a$ on a smooth surface, subjected to an impulse, J, at one end. G is the centre of mass of the stick.

The relationship of v and ω is $\omega = 3v/a$ or $v = \frac{1}{3}\omega a$. This means that the rotational motion is centred at a point, say O, distant $\frac{2}{3}a$ from the end of the stick at which the impulse made its impact. The motion is a combination of a translational velocity, $v = 2\,\mathrm{m\,s}^{-1}$, of the centre of mass and a rotation of angular velocity, $\omega = 12\,\mathrm{s}^{-1}$, about the centre of mass.

Further examination of this result brings to light two interesting points. If the boy's kick were to pass through the centre of mass of the stick, its motion would be only a linear motion of the centre of mass. The kinetic energy of the stick would then be $T = J^2/2m$. When the kick is directed at the end of the stick, the kinetic energy of the stick is the sum of its kinetic energies of translation and of rotation, $T = J^2/2m + 3J^2/2m = 2J^2/m$. The same impulse applied to different points produces different kinetic energies. There is a ready explanation for this if we assume that the impulse is a constant force, F, applied for a time t, $J = Ft$. This impulse applied to the centre of mass produces a translational acceleration F/m and at time t the centre of mass of the stick will have travelled through a distance of $\frac{1}{2}(F/m)t^2$. The work done by the force is then $(F^2/2m)t^2 = J^2/2m$. When the impulse is applied to the end of the stick, it produces an angular acceleration Fa/I and at time t the end of the stick will have moved through an angle θ, with $\theta = \frac{1}{2}(Fa/I)t^2$. With $I = \frac{1}{3}ma^2$, $\theta = 3Ft^2/2ma$, so that the distance travelled by the end of the stick because of its rotation is $a\theta = 3Ft^2/2m$ and the work done by the force is $Fa\theta = 3J^2/2m$. The work done by the force, the sum of the work done in producing the linear and angular displacements, is four times greater than when the impulse is applied to the centre of mass. This description will be valid as long as the distance $a\theta$ is effectively in the same direction as that of the force, so that we can identify $a\theta$ as being the distance moved by the force in its own direction. The time interval t has to be a small time interval, just as would be expected when the impulse is produced by a kick.

The second point is that we have used the equality of the angular impulse and the change in angular momentum about the centre of mass to calculate ω. The angular momentum about G is simply Ja. On the other hand, the moment of inertia about the (initial) centre of rotation, O, is $m[\frac{1}{3}a^2 + (\frac{1}{3}a)^2] = 4ma^2/9$, and the angular momentum about O is $I_O\,\omega = 4Ja/3$. The difference between the two angular momenta is simply $\frac{1}{3}mva$, the moment of the linear momentum of the centre of mass about O. When G is our reference point, the moment of momentum is zero.

As a further example of constructing angular momentum from the moment of a linear momentum and an angular momentum with respect to the centre of mass, revisit Example 10.5. In that case we could say that the angular momentum of the rolling sphere with respect to its point of contact, C, with the slope was $I_C\,\omega$, where I_C is the moment of inertia about C. If the sphere radius is a, then $I_C\,\omega = \frac{7}{5}Ma^2\omega$. The angular momentum with respect to the centre of mass, G, is written as $I_G\,\omega = \frac{2}{5}Ma^2\omega$. The difference between these two angular momenta is the moment of momentum of the sphere about G. This is $Mva = M\omega a^2$.

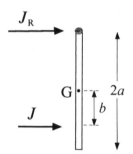

Fig. 10.13 The impulses acting on a pub sign suspended from its top edge when a snowball strikes it.

10.8. This example has similarities with Example 10.7, but this time the rotational motion is about the hinge rather than about some point in the rotating body. This change must be due to the effect of a reactive impulse, say J_R, at the hinge additional to the impulse, J, of the snowball as it strikes the sign. The existence of J_R leads to the situation shown in Fig 10.13, in which the sign has a length $2a$ and the snowball impinges at a distance b below the centre of mass, G, of the sign. The total impulse acting on the system is $J + J_R$, so that

$$J + J_R = Mv,$$

where Mv is the change in momentum of the sign and, taking G as reference point,

$$bJ - aJ_R = I_G \omega.$$

Rotation about the hinge leads to the identity $v = a\omega$, and for the rectangular sign $I_G = \frac{1}{3}Ma^2$, so that the angular momentum equation becomes

$$bJ - aJ_R = \frac{1}{3}Ma^2(v/a) = \frac{1}{3}Mav.$$

Evaluating J_R from the two equations in the J's and v leads to

$$J_R = J\frac{b - \frac{1}{3}a}{a + \frac{1}{3}a} = \frac{3}{4}(J/a)(b - \frac{1}{3}a).$$

Assuming that the mass of the snowball is m and its velocity V is reduced to zero by its impact on the sign, then

$$J = \delta(mV) = 4.44 \, \text{Ns}.$$

Since $a = 0.75$, $b = 0.55$, and $J_R = 1.33 \, \text{Ns}$. The angular velocity of the sign may be calculated from either of the impulse equations as $0.64 \, \text{s}^{-1}$.

The expression defining the reactive impulse J_R tells us that, for $b > \frac{1}{3}a$, J_R is in the same direction as J, while for $b < \frac{1}{3}a$ the two impulses are oppositely directed. When $b = \frac{1}{3}a$ the value of J_R is zero; there is no impulse at the hinge. The point of impact for which $J_R = 0$ is called the centre of percussion of the sign, referred to the position of the hinge. If a cricket ball strikes a cricket bat at its centre of percussion, referred to the 'point' at which it is held by the batsman, the batsman feels no sensation of impact in his hands and arms. In the design of hammers and other percussion tools, as for sports racquets and bats, the relationship between head position, centre of mass position, and the position of the centre of percussion is an important consideration.

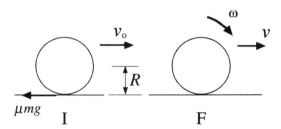

Fig. 10.14 The skittle ball of radius R moving in its initial state I with velocity v_0 and in its final state F with velocity v and angular velocity ω. The friction force μmg acts through the point of contact between ball and floor.

10.9. As shown in Fig. 10.14, the only force acting on the sliding ball is due to friction at the point of contact of the ball with the floor. This is the product of the contact force and the coefficient of friction which, for a ball of weight W and a coefficient of friction, μ, is $\mu W = \mu mg$. The force acts to reduce the velocity of the ball so that

$$m(dv/dt) = -\mu mg$$

or

$$dv/dt = -\mu g.$$

If v_0 is the velocity of projection, then at any time t after projection the velocity v is given by

$$v = v_0 - \mu gt.$$

The torque (referred to the centre of mass) acting on the ball is μmgR, so that the equation of rotational motion is

$$\mu mgR = I(d\omega/dt) = \tfrac{2}{5}mR^2(d\omega/dt),$$

leading to $\omega = \omega_0 + (5\mu g/2R)t$ and, since $\omega_0 = 0$, $\omega = (5\mu g/2R)t$. When the ball begins to roll $v = \omega R$ and so

$$\tfrac{5}{2}\mu gt = v_0 - \mu gt \qquad \text{or} \qquad t = 2v_0/7\mu g.$$

We assume that rolling without slipping is a non-dissipative process, so that the velocity of the ball when it reaches the skittle pins will be $6.43\,\text{ms}^{-1}$.

The velocity of the ball when rolling without slipping starts is $5v_0/7$ and the distance travelled is $x = v_0 t - \tfrac{1}{2}\mu gt^2 = 12v_0^2/49\mu g = 10.12\,\text{m}$.

When the ball has initial rotation, let's say that its initial angular velocity is ω_0. Then $\omega - \omega_0 = (5\mu g/2R)t$ so that $R(\omega - \omega_0) = \tfrac{5}{2}(v - v_0)$

or $v = \frac{5}{7}(v_0 + R\omega_0)$, where v_0 and $R\omega_0$ are unlikely to be equal. What is noticeable is that if ω_0 is negative, $v < \frac{5}{7}v_0$, while if $\omega_0 > 0$, $v > \frac{5}{7}v_0$. This implies, in this example, that $\omega_0 < 0$ for backspin and $\omega_0 > 0$ for forward spin (rotating in the direction of linear motion).

The solution of this example can be simplified by considering the angular momentum of the ball relative to its point of contact with the floor. Because the friction force passes through this point we have a torque-free system and angular momentum is conserved. In this case the initial angular momentum and the final angular momentum must be equal, so that the initial angular momentum of the centre of mass is the sum of the final angular momentum of the centre of mass and the angular momentum of the sphere rotating about its centre,

$$mv_0 R = mvR + I\omega$$

and, since $v = R\omega$ in the final state,

$$v = v_0/(1 + I/mR^2)$$

which, for a solid sphere with $I = \frac{2}{5}mR^2$, is $\frac{5}{7}v_0$.

It's of interest to note that the friction force does the same quantity of work independent of the value of μ.

11
Rotational motion—angular momentum I

11.1 Vector angular momentum

So far, we have dealt with rotational motion in a scalar representation. As an example, we have said that a tangential force, F, applied to a rod at a distance r from a pivot, produces a torque $\Gamma = Fr$. What happens when the force is not tangential, but is applied in a direction inclined at an angle θ with respect to the rod, is shown in Fig. 11.1. The force can be resolved into its components, a radial component $F \cos \theta$ directed along the rod towards the pivot point, O, and a tangential component $F \sin \theta$ perpendicular to the length of the rod. Obviously, the radial component tends only to compress the rod and makes no contribution to the rotational motion. The torque which gives rise to the rotation is due to the tangential component of the force and is the product of $F \sin \theta$ and r,

$$\Gamma = Fr \sin \theta.$$

If Γ were represented by a vector $\boldsymbol{\Gamma}$, the vector would have a magnitude $|\boldsymbol{\Gamma}| = \Gamma = rF \sin \theta$, where θ is the angle between the directions of $\boldsymbol{r}$ and $\boldsymbol{F}$. Since r and F are the magnitudes of vectors, and since $\boldsymbol{\Gamma}$ has the appropriate magnitude, this leads us to propose that $\boldsymbol{\Gamma}$ is the vector product of $\boldsymbol{r}$ and $\boldsymbol{F}$, i.e.

$$\boldsymbol{\Gamma} = \boldsymbol{r} \times \boldsymbol{F},$$

in which case, $\boldsymbol{\Gamma}$ is directed perpendicularly upwards out of the plane of Fig. 11.1, as illustrated in Fig. 11.2, from which it is seen that $\boldsymbol{\Gamma}$ is perpendicular to the plane containing $\boldsymbol{r}$ and $\boldsymbol{F}$. Since $\boldsymbol{\Gamma} \propto d\boldsymbol{\omega}/dt$, a vector representing angular acceleration would be parallel to $\boldsymbol{\Gamma}$. We would then expect that a vector representing angular velocity, $\boldsymbol{\omega}$, would be parallel to $\boldsymbol{\Gamma}$ when the torque is an accelerating torque and antiparallel to $\boldsymbol{\Gamma}$ when the torque is a retarding one. $\boldsymbol{\Gamma}$ and $\boldsymbol{r}$ are always in directions perpendicular to one another, and so $\boldsymbol{\omega}$ is always perpendicular to $\boldsymbol{r}$, as shown in Fig. 11.3. The tangential velocity around the arc of radius r has a magnitude $r\omega$, and, because it is always directed perpendicular to $\boldsymbol{r}$, may be written as the vector

$$\boldsymbol{v} = \boldsymbol{\omega} \times \boldsymbol{r},$$

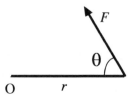

Fig. 11.1 A force F with its direction inclined at an angle θ to the length of a rod which is free to turn about a pivot at O.

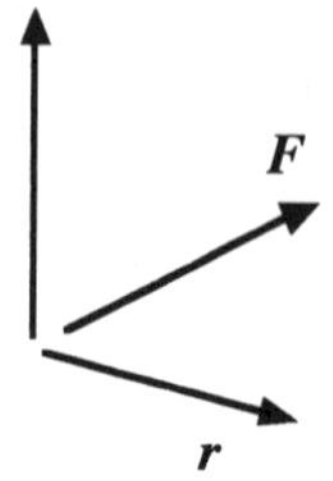

Fig. 11.2 The vector relation between F, r, and Γ.

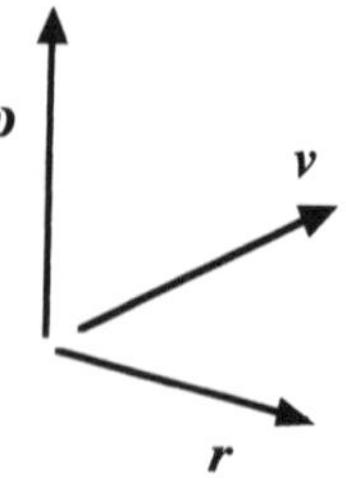

Fig. 11.3 The vector relationship of ω, r, and tangential velocity, v.

because the angle between r and ω is always 90°. Angular momentum as a vector quantity must be parallel to ω (and parallel to Γ for an accelerating torque, or antiparallel for a retarding one). This means, for example, for a particle of mass m travelling in a path of radius r with an angular velocity ω, that the angular momentum of the particle is

$$l = m(r \times \omega \times r) = r \times mv = r \times p,$$

where p is the linear momentum of the particle, once again emphasizing that angular momentum is the moment of momentum.

If l has components l_x, l_y and l_z, in the x-, y-, and z-directions, these components can be written in form involving the components of r and p, as

$$l_x = yp_x - xp_y, \qquad l_y = zp_x - xp_z, \qquad \text{and} \qquad l_z = xp_y - yp_x,$$

since, for unit vectors, $|i \times i| = 0$, $|j \times j| = 0$, and so on. Furthermore, as would be expected,

$$|l| = (l \cdot l)^{1/2} = \left(l_x^2 + l_y^2 + l_z^2\right)^{1/2}.$$

We equate angular impulse with the change in angular momentum so that angular impulse is a vector quantity. On the other hand work, kinetic energy, and power in rotational motion are scalar quantities, as indeed they must be.

11.2 Total angular momentum

Example 62. A bowler in a bowling alley watches his bowling ball rolling towards the pins. If he could see the vector representing the angular momentum of the ball, in which direction would he see it point? What difference would it make if he imparted to the ball a spin about its vertical axis as he launched it?

As the ball, of radius R, moves away from the bowler it attains the state in which $v = R\omega$. The ball is then spinning, so that $\omega \times R = v$. Since v is directed away from the bowler, ω and R are arranged perpendicular to each other and are both perpendicular to v. R joins the centre of the ball to the point of contact with the floor; that is, it is vertical, so that ω is horizontal and points to the bowler's left: l is parallel to ω. Alternatively, the direction of l could have been derived from $l = R \times p$ or, more adventurously, from $l = m(R \times \omega \times R)$.

We can deal with the case in which the ball is spinning about its vertical axis as its rolls along by identifying the rolling angular momentum with a component l_y and the angular momentum due to the spin about the vertical axis as a component angular momentum, l_z. The total angular momentum due to the spinning of the ball about the y- and z-axes will then have a magnitude $|l| = (l_y^2 + l_z^2)^{1/2}$ and will be inclined at an angle of $\tan^{-1}(l_z/l_y)$ to the horizontal (the y-axis): l_z will be negative or positive depending on whether the spin is left-handed or right-handed (anti-clockwise or clockwise when viewed from above). Within the limitations we have imposed, that is, when the rolling motion is unaffected by frictional forces, there is no torque acting on the ball so that the angular momentum vector will remain fixed in magnitude and direction. In a real situation the point of contact will be an area of contact rather than a point. There will be dissipative effects and the rolling ball won't continue rolling forever.

Here we have formed a total angular momentum for the system from its component angular momenta. The two angular momenta due to two spinning motions of the ball have been combined by vector addition. In Example 10.9 we added the angular momentum due to the spin of the ball to the angular momentum due to linear motion by using the point of contact of the ball with the floor as reference point. We were enabled to do that because the two angular momenta were collinear and could be added simply as scalar quantities. In general, the addition of angular momenta is an exercise in vector addition.

Example 63. The force acting on a planet in the solar system to keep it in its orbital path is directed towards the sun, which is distant $r = |r|$ from the planet. If the earth has a mass of 5.98×10^{24} kg and takes 1 year to complete a circuit of its orbit, show that the orbital angular momentum of the earth is constant and that the line which joins the earth to the sun sweeps out equal areas in equal times. What is the magnitude of the earth's orbital angular momentum if its orbit may be approximated as a circle of radius 1.5×10^{11} m?

The force acting on the planet earth may be taken as F, and since F is directed towards the sun F and r are collinear. This means that the vector product, $r \times F$, of r and F is zero. There is no torque acting on the system, or $\Gamma = 0$. Since $\Gamma = dl/dt = 0$, the angular momentum doesn't change with time: l is constant, whatever the position of the planet in its orbit. In a time interval δt, when the angular velocity of the earth is ω, the radius vector r alters its angular position through an angle $\omega \, \delta t$, as shown in Fig. 11.4. The area of the triangle swept out by the vector is

$$\delta A = \tfrac{1}{2} r \times \omega \, \delta t \times r.$$

In scalar terms, $\delta A = \tfrac{1}{2} r^2 \omega \, \delta t$, or

$$dA/dt = l/2m.$$

Since both l and m are constant, dA/dt is constant, the line joining the earth to the sun sweeps out equal areas in equal time intervals. Because dA/dt is constant we can identify its magnitude of dA/dt with the area of the orbit divided by the

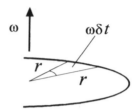

Fig. 11.4 The triangular form produced by the line joining the earth to the sun as the earth progresses round its orbit.

time taken for the earth to complete one orbit; that is, $dA/dt = A/\mathscr{T}$. $\mathscr{T}$ is, of course, 1 year or so

$$|l| = 2m\,\frac{dA}{dt} = \frac{2mA}{\mathscr{T}} = 2.68 \times 10^{40}\ \mathrm{N\,s\,m}.$$

The orientation of the angular momentum vector is perpendicular to the plane of the orbit.

Since Kepler's work early in the seventeenth century, it has been well known that the earth's orbit is an ellipse rather than a circle, and yet we have assumed that the orbit is circular. In justification, we make the observation that the orbit, though elliptical, is not very different in shape from a circle. The area we have calculated for the circular orbit differs from the area of the actual elliptical orbit only by around 0.014%. This difference is more or less the same for the other planets which were known in the seventeenth century, giving an indication of the precision of Tycho Brahe's measurements, on which Kepler based his calculations.

One point that we have overlooked in examining the angular momentum of the earth is its rotation about its own axis. The axis is inclined at an angle of 66.5° to the plane of the orbit, and the earth completes one revolution about the axis in a day. If we take the radius of the earth to be $a = 6.37 \times 10^6$ m we can calculate this angular momentum as $\frac{2}{5}ma^2\Omega^2$ (with $\Omega = 2\pi$ per day), giving a value of about 7×10^{33} kg m^2 s^{-1}. Since this is so small compared with the orbital angular momentum of the earth, adding the earth's spin angular momentum on to its orbital angular momentum will produce a total angular momentum that is effectively the same as the orbital angular momentum in magnitude and in direction.

So far, the central force, acting along the line which joins planet to sun, hasn't been specified in detail, so that the constant value of the angular momentum and the area rule should apply independent of the function which defines F in terms of r. On the other hand, the planets continue to travel around their orbits, which must be taken as bound states of the planets–sun system. Such bound states have negative total energy values. The total energy is the sum of the kinetic and potential energies,

$$H = T + V = \tfrac{1}{2}mr^2\omega^2 + V,$$

so that for H to be negative $V < 0$, because $T > 0$. Taking the form of F to be

$F = -C/r^n$ gives $V = -C/[(n-1)r^{n-1}]$. The magnitude of the central force must be equal to the centrifugal force,

$$C/r^n = mr\omega^2,$$

leading to

$$C/r^{n-1} = mr^2\omega^2,$$

so that

$$H = \tfrac{1}{2}mr^2\omega^2 - [mr^2\omega^2/(n-1)]$$
$$= mr^2\omega^2[(n-3)/(2n-2)].$$

For $n = 1$, $H = -\infty$, which is impossible, while for $n \geqslant 3$, $H \geqslant 0$, so that only if $1 < n < 3$ can there be bound states. If we take a reductionist approach by saying that n must be an integer, then only $n = 2$ gives rise to bound states. This is the form of the gravitational law proposed by Newton.

Exercises

11.1. A pool ball has a diameter of 4 cm. It is set into motion by a sharp, horizontal, impulse provided by a cue. At what height above the table should the cue strike the ball so that the subsequent motion is rolling without slipping?

11.2. Two circular discs with rough surfaces can rotate about the same axis perpendicular to their planes, passing through their centres. Initially the discs are not in contact; one disc with moment of inertia I_1 rotates with angular velocity ω_1 and the other, with moment of inertia I_2, is static. If the discs are brought into contact, what is the angular velocity of the combination of the two discs in contact? What energy loss occurs when they are brought into contact?

11.3. Two astronauts involved in a satellite repair mission have got themselves into a situation in which they are at opposite ends of a light girder of length 10 m. The astronauts have a mass of 75 kg apiece and their velocity with respect to the centre of mass of the astronaut/girder system is $5 \, \text{ms}^{-1}$. By how much do their velocities change if they haul themselves along the girder until they are 5 m apart? How much work have they done in moving closer together?

11.4. A figure skater whose mass is 60 kg begins a spin with her arms extended horizontally from her shoulders. When her rate of rotation is 30 rpm she allows her arms to drop to her sides. What is her new rate of spinning? (Refer to the table in Fig. 4.7 for anatomical data).

11.5. A half-empty spool of cotton thread lies at rest on a rough horizontal floor. The diameter of the spool is 3 cm and it is filled with thread to a diameter of 2.7 cm. A child takes hold of the loose end of the thread and pulls it in a direction inclined at an angle of 36° above the horizontal. Will the spool roll in the direction in which the thread is pulled?

11.6. A body of mass m travels in a closed orbit under the influence of a central force $F(r) = -C/r^2$, where C is a constant and r is the separation of the body from the centre of force. What is the relationship between the energy of this system and the angular momentum of the body?

11.7. The engine of a helicopter produces 1000 hp when it hovers in still air. The main rotor sweeps out a circle of radius 5 m. Make an estimate of the force which must be produced by the tail rotor situated 10 m behind the axis of rotation of the main rotor to prevent the helicopter from rotating about a vertical axis.

Solutions

11.1. The pool ball is illustrated in Fig. 11.5, in which the impulse, J, is applied at a height h above the table: h may be defined in terms of the radius, a, of the ball as

$$h = a(1 - \cos \theta),$$

where θ is the angle between the diameter passing through the point of contact (of the ball and the table) and the radius passing through the point of impact. The impulse, J, is equal to the change in linear momentum of the ball. If the mass of the ball is m, the velocity imparted to the ball is $v = J/m$. The angular impulse is $J(h - a) = -Ja \cos \theta$, and is equal to the change in angular momentum, $I\omega$. For the solid spherical ball $I = \frac{2}{5}ma^2$, so that $\omega = -5J \cos \theta/2ma$. For rolling without slipping, $v = a\omega$; i.e.,

$$J/m = -\tfrac{5}{2}(J/m)\cos \theta, \qquad \text{or} \qquad \cos \theta = -\tfrac{2}{5},$$

so that

$$h(1 - \cos \theta) = \tfrac{7}{5}a = 2.8 \text{ cm}.$$

If the ball is struck at this height it suffers no energy loss due to reduction either in v or in ω: $\frac{7}{5}a$ is the position of the centre of percussion of the ball relative to the surface of the table. The impulse at the point of contact between the ball and the table is zero. When this contact impulse is non-zero the ball skids along the surface because $v \neq a\omega$. In reality, it seems unlikely that the cue will deliver a horizontal impulse to the ball. Then the linear momentum will be equal to the horizontal component of the impulse and the angular impulse will be a combination of the angular impulses due

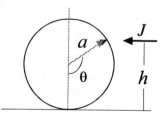

Fig. 11.5 The pool ball being subjected to an impulse, J, at a height h above the table.

to the vertical and horizontal components of the impulse. The vertical component of the impulse makes it possible for the ball to develop a backward velocity at some time after the cue has struck the ball.

11.2. The initial angular momentum of the two-disc system is that of the rotating disc, $I_1 \omega_1 = L_1$. After contact is made, because the surfaces are rough, the two discs will attain a state in which they rotate together with the same angular velocity, ω_f. Then the angular momentum will be $L_f = (I_1 + I_2)\omega_f$. There is no external torque acting, and so

$$L_1 = L_f, \quad \text{or} \quad I_1 \omega_1 = (I_1 + I_2)\omega_f,$$

giving

$$\omega_f = \frac{I_1}{(I_1 + I_2)} \omega_1.$$

Initially, the rotational energy is $T_1 = \frac{1}{2}I\omega_1^2$, and in the final state it is

$$T_f = \frac{1}{2}(I_1 + I_2)\omega_f^2$$

which, in terms of ω_1, is

$$T_f = \frac{1}{2}(I_1 + I_2)\frac{I_1^2}{(I_1 + I_2)^2}\omega_1^2.$$

The loss of energy is

$$T_1 - T_f = \frac{1}{2}I_1\omega_1^2\left(\frac{I_2}{I_1 + I_2}\right) = T_1\left(\frac{I_2}{I_1 + I_2}\right).$$

The energy of the combined discs could be written as

$$T_f = T_1\left(\frac{I_1}{I_1 + I_2}\right).$$

In this situation the energy loss which occurs is the work done in accelerating the static disc to an angular velocity of ω_f.

If we extend this problem so that both discs have initial angular velocities, say ω_1 and ω_2, a similar argument shows that the energy loss is

$$\delta T = \frac{1}{2}\left(\frac{I_1 I_2}{I_1 + I_2}\right)(\omega_1^2 - \omega_2^2),$$

which is zero for $\omega_1 = \omega_2$.

These two discs, which can be separated or in contact, might be seen as the simplest analogue of the clutch plates in an automobile with a manual gearbox. Then, I_1 and I_2 would be the effective moments of inertia of the engine–drive plate combination and the gear train–transmission–wheels–vehicle combination. When the clutch plates are brought into contact the energy loss is compensated by the engine. If the plates come together very quickly, the rate of energy loss may be more than the power available from the engine, so that the engine stalls.

11.3. Both astronauts are travelling in a circular path of radius 5 m and their angular velocities are simply $\omega = v/r = 1\,\text{s}^{-1}$. Provided that we can treat their masses as point masses (an oversimplification), that is, their centres of mass effectively coincide with the ends of the girder and they don't move their limbs about, we can say that their initial angular momentum is $L = mr^2\omega = 150 \times 25 = 3750\,\text{N m s}$. Since they move radially inwards towards the centre of the system, they exert no torque as they move in and angular momentum is conserved. Their new value of angular velocity is obtained from $150 \times 2.5^2 \times \omega = 3750$, giving $\omega = 4\,\text{s}^{-1}$, so that their new velocities are $10\,\text{m s}^{-1}$. Their velocities have increased by a factor of two.

The work done is found most simply from the change in kinetic energy that has taken place:

$$W = \tfrac{1}{2}mr_2^2\omega_2^2 - \tfrac{1}{2}mr_1^2\omega_1^2 = 5625\,\text{J}.$$

It's not quite so easy to calculate the work done by multiplying force by distance moved in the direction of the force, but it isn't difficult either. The force acting on the rotating astronauts as they whirl around the centre of mass is the centrifugal force $mr\omega^2$. In pulling themselves in by a short distance δr they perform work $\delta W = -mr\omega^2\,\delta r$. The work done is found by integrating δW from $r = 5$ to $r = 2.5$. Then we recall that both r and ω are variable quantities, so that we need an alternative expression for the force in order to perform the integration. This is obtained readily with the observation that angular momentum is conserved, so that $L = 3750\,\text{N m s}$ is constant. This enables us to write the centrifugal force as

$$F = mr\omega^2 = \frac{(mr^2\omega)^2}{mr^3} = \frac{L^2}{mr^3},$$

so that

$$W = -(3750^2/150)\int_5^{2.5} r^{-3}\,dr$$

$$= 5625\,\text{J}.$$

When angular momentum is conserved in an orbital motion the centrifugal force may be written in the form

$$F = L^2/mr^3.$$

11.4. Assuming that no torque acts while the skater lowers her arms, angular momentum is conserved: $I_1\omega_1 = I_2\omega_2$, where I_1 and I_2 are the skater's moments of inertia with arms extended and to her sides respectively, and $\omega_1 = 30\,\text{rpm} = \pi\,\text{s}^{-1}$. We have to make estimates of the ratio of the moments of inertia in order to find the new rate of rotation.

The change that makes I_2 different from I_1 is an arm movement only, so that the contribution to the moment of inertia from the skater's head, torso, and legs, I_b, is constant. $I_1 = I_b + I_e$ and $I_2 = I_b + I_s$, where I_e and I_s are the moments of inertia of the arms when extended and to her sides respectively. Take I_b as the moment of inertia of a rectangular block, of sides $2a$ and $2b$ and height h, about and axis parallel to h through its centre of mass. Then $I_b = \frac{1}{3}M_b(a^2 + b^2)$, where M_b is the mass of the skater's head, torso, and legs. From the table of masses and position of the centres of mass of the body's components (in Fig. 4.7), $a = 0.107 \times 1.7 = 0.182$ m. We may estimate b as 10 cm, and $M_b = 0.875 \times 60 = 52.5$ kg. This gives $I_b = 0.755$ kg m². Because the detailed distribution of mass in the skater's arms is not specified, it is easiest to treat the arms as a combination of point masses located at the centre of mass of the upper arm, the forearm, and the hand. Again, from the table, the centre of mass of the upper arm is situated at a distance of $(0.8116 - 0.7174) \times 1.7$ m from the shoulder joint, so that when the arms are extended the centre of mass of the upper arms are $(0.0942 + 0.107) \times 1.7 = 0.342$ m from the axis of rotation. Similarly, for the forearm and the hand the centres of mass are at distances of 0.621 m and 0.828 m from the axis of rotation. The mass of both upper arms is $0.066 \times 60 = 3.96$ kg, and those of the forearms and hands are 2.52 kg and 1.02 kg respectively, so that

$$I_e = 3.96 \times 0.342^2 + 2.52 \times 0.621^2 + 1.06 \times 0.828^2 = 2.162 \text{ kg m}^2.$$

When the arms are held to the skater's sides, the centres of mass all lie at the same distance from the axis of rotation, 0.182 m. so that

$$I_s = 7.54 \times 0.182^2 = 0.25 \text{ kg m}^2.$$

Thus we have $I_1 = 2.917$ kg m² and $I_2 = 1.0$ kg m². The new rate of rotation is

$$\omega_2 = I_1 \omega_1 / I_2 = 2.92\pi \text{ s}^{-1}$$

or nearly 88 rpm. To all intents, the rate of rotation has trebled.

We can make an estimate for an upper limit to the rate of rotation by saying that the limiting case occurs when the arm and hand centres of mass are situated on the axis of rotation, a slight exaggeration of what is anatomically possible. In this case $I_2 = I_b$ and the skater would spin at 3.87 times the original rate.

11.5. When the thread is pulled, the contact force, F_C, isn't normal to the plane of the surface on which the spool rests, but has a component parallel to the surface. The forces acting on the spool are shown in Fig. 11.6. There are two

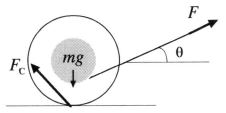

Fig. 11.6 The forces acting when the thread on a spool is pulled.

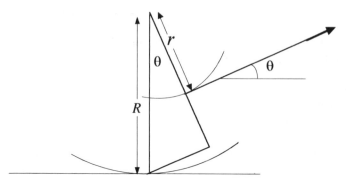

Fig. 11.7 The geometry used to establish the torque acting on the spool when the thread is pulled.

obvious axes parallel to the spool length which might be used to establish torques and angular acceleration of the spool, one passing through the point of contact between the spool and the surface and the other passing through the centre of mass of the spool. The advantage of using the point of contact as axis is that the contact force passes through the axis and produces no torque. We don't need to know anything about F_c. The value of the torque acting on the spool in this case is established as shown in Fig. 11.7, whence it is seen that

$$\Gamma = F(R \cos \theta - r),$$

where R is the spool radius and r is the outer radius of the thread wound on the spool. Then, if I is the moment of inertia of the spool about the axis passing through the point of contact, we have

$$\Gamma = F(R \cos \theta - r) = I(\mathrm{d}\omega/\mathrm{d}t) = (I_G + mr^2)(\mathrm{d}w/\mathrm{d}t),$$

where I_G is the moment of inertia about the axis passing through the centre of mass and m is the mass of the spool and cotton.

The angular acceleration, $\mathrm{d}\omega/\mathrm{d}t$, is greater than or less than 0 depending on whether $R \cos \theta$ is greater than or less than r. Because the surface is defined as rough, we may assume that $\mathrm{d}v/\mathrm{d}t = R(\mathrm{d}\omega/\mathrm{d}t)$, so that the linear acceleration will be positive for $\cos \theta > r/R$; in other words, for angles $\theta < \cos^{-1}(r/R)$ the spool will move in the direction of the horizontal component of the applied force. When $\theta > \cos^{-1}(r/R)$ the spool will move away from the child who is providing the force. We may define a critical angle $\theta_0 = \cos^{-1}(r/R)$ which separates these two regimes in which the force has opposite. If the force is applied along the direction defined by θ_0 the spool doesn't move away from its original position. In this particular case $r/R = 2.7/3 = 0.9$ and $\theta_0 = 25.8°$. The child pulls with $\theta = 36°$, so that the spool moves away from him.

Had we taken the centre of mass axis the torque would have taken the form $\Gamma = RF_{cx} - Fr$, where F_{cx} is the horizontal component of F_c, giving

$$F_{cx}R - Fr = I_G(\mathrm{d}\omega/\mathrm{d}t).$$

The equation for the linear acceleration would have been

$$F \cos \theta - F_{cx} = m(dv/dt) = mR(d\omega/dt).$$

Eliminating F_{cx} from these two equations leads to an expression for the angular acceleration which is no different from the one given above. The extra complication brought about by having to use the equation of linear motion makes the introduction of error more probable than when the axis is chosen to pass through the point of contact of the spool and the surface.

11.6. The energy of the system is the sum of the kinetic and potential energies of the body as it travels in its orbit,

$$H = T + V(r)$$

and, since the force is related to the potential by $F(r) = -dV/dr$,

$$V(r) = -C/r = rF(r).$$

The angular momentum of the body $L = mr^2\omega$. The body following its orbital path is in an equilibrium state, so that it will have a nett force of zero acting on it. The centrifugal force and the central force are equal in magnitude and opposite in sign so that $F(r) + mr^2\omega = 0$, or

$$F(r) = -L^2/mr^3 = -C/r^2.$$

This means that we can write $V(r) = rF(r) = -L^2/mr^2$. Then the energy of the system is

$$H = \tfrac{1}{2}mr^2\omega^2 - L^2/mr^2$$

and, since $\tfrac{1}{2}mr^2\omega^2 = \tfrac{1}{2}(L^2/mr^2)$,

$$H = -L^2/2mr^2.$$

From the relation between $F(r)$ and L, we have that $r = L^2/Cm$, so that, eliminating r from the expression for H,

$$H = -\frac{L^2C^2m^2}{2mL^4} = -\frac{C^2m}{2L^2}.$$

If we took the central force to be electostatic, between two charges $-e$ and Ze, then

$$C^2 = \left[Z^2e^4/16\pi^2\varepsilon_0^2\right]$$

(with ε_0 the permittivity of free space), and the expression for the total energy would take the form

$$H = -\frac{Z^2e^4m}{32\pi^2\varepsilon_0^2L^2}$$

so that, if we choose that L is quantized in units of $h/2\pi$, that is, $L = n(h/2\pi)$, with n an integer and h a constant, Planck's constant,

$$H = -\frac{Z^2e^2m}{8\varepsilon_0^2n^2h^2}$$

which, if e and m are electronic charge and mass respectively, is the equation describing the energy states of an electron in a one electron atom with atomic number Z in terms of a single quantum number n, Z, and a set of constants. For the hydrogen atom $Z = 1$, for the singly charged helium ion He^+ $Z = 2$, and for the doubly charged lithium ion Li^{++} $Z = 3$. Often $h/2\pi$ is written as $\hbar$ giving the equation for the energy of the energy states as $H = -Z^2 e^4 m / 32\pi^2 \varepsilon_0^2 \hbar^2$ (with ε_0 the permittivity of free space). This equation for H is the central result of the Bohr theory of the atom.

11.7. Power is the rate of working, so that for work W performed in time t, $P = W/t$. If the torque, Γ, producing the rotation gives a rotation of θ in time t, then $P = \Gamma\theta/t$. When the angular velocity, ω, is constant—as it is in this case—$\omega = \theta/t$ and $P = \Gamma\omega$. Since 1 hp = 746 W, the torque produced as the main rotor spins is $\Gamma = P/\omega = (7.46 \times 10^5)/(40\pi) = 5.94 \times 10^3$ N m. The tail rotor lies 10 m from the centre of the main rotor and must produce a torque of equal magnitude if it is to prevent rotation of the helicopter fuselage. The force required is simply $F = 0.1 \times \Gamma = 594$ N.

If we knew the mass of the helicopter, say 10 tonnes, and the desirable rate of rotation of the tail rotor, say 3000 rpm, we could make a crude estimate of the diameter of the tail rotor by saying that the rotors produce their forces by each displacing a cylinder of air which has the same radius as the rotor. The rate at which momentum is given to the air is the force produced by the rotor. There are analogues to the calculation of the force produced by a water jet which was examined in Example 18. The force exerted by the water was $\rho A v^2$, where ρ is the density, A is the cross-sectional area, and v is the velocity of the water jet. Assuming we can use the same expression when the displaced fluid is air (despite the fact that water is an incompressible fluid and air is a compressible fluid) we are left to decide on the appropriate value of v. At the rotor blade tips this would be proportional to the rotor radius (the length of a rotor blade), R, at the centre it would be zero. The average value, because v is linear in r, is $\frac{1}{2}\omega R$. Since the area of the jet is πR^2 we have $F \propto R^4 \omega^2$. For the main rotor and the tail rotor we may form the ratio

$$\frac{F_m}{F_t} = \frac{R_m^2 \omega_m^2}{R_t^4 \omega_t^2} = (5 \times 1600\pi^2)/(10\,000\pi^2 \times R_t^4)$$

Then $F_m = mg = 9.8 \times 10^4$ N and $F_t = 594$ N; so that $R_t = 0.88$ m.

It's quite remarkable that this value is in the right ball park (of the right order of magnitude) despite the simplistic nature of the calculation.

12

Rotational motion—angular momentum II

12.1 Polar equations of motion

Often we are told that the modern physical sciences had their origin in the work of Niklaus Copernicus. In the middle of the sixteenth century Copernicus proposed that the planets in the solar system travelled in circular orbits with the sun situated at their centres. Precise measurements made by Tycho Brahe within the next fifty years showed that this theory was untenable. Kepler, after years of calculation and speculation about those observations, demonstrated that the orbits were elliptical, with the sun situated at a focus of the ellipse, not circular with the sun at the circle centre. The distance of a planet from the sun varies with time and the shape of the orbit is no longer defined by a single radius. If we are to try to explain the elliptical shape of planetary orbits we need a technique which is slightly more advanced than those we have used previously.

It was not until about fifty years after Kepler published his ideas that Newton developed the mathematics which enabled him to astound the world by producing his theory of universal gravitation to explain the motions of the planets.

Example 64. A particle of mass m is travelling in a curved path under the influence of a central force, $F(r)$. What are its linear accelerations along the line which joins the particle to the centre of force and in a direction perpendicular to that line?

So far, we have used the principle of conservation of momentum to duck the question of providing equations of motion for a particle travelling along a path of variable curvature. Here we are asked to establish the components of the equation of motion, for example, for a planet moving under the influence of the gravitational force when the separation of the planet from the centre of force is variable in time. It's convenient to describe the position of the planet by the position of its centre of mass; that is, to treat the planet as if it were a point particle. The best choice of coordinate system to describe the position of the particle is polar coordinates, r and θ, the origin of which is situated at the centre of force. Then we say that the particle has a radial velocity $v_r = \mathrm{d}r/\mathrm{d}t$ and an angular velocity $\omega = \mathrm{d}\theta/\mathrm{d}t$, as shown in Fig. 12.1. The linear velocity of the particle in a direction perpendicular to the direction of the radius line is

$$v_\theta = r\omega = r\,\mathrm{d}\theta/\mathrm{d}t.$$

It's only when the radius of curvature of the particle path is constant that v_θ can be identified with the tangential velocity v_t; otherwise,

$$v_t^2 = v_r^2 + v_\theta^2.$$

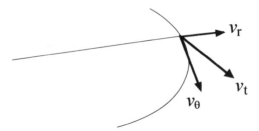

Fig. 12.1 The components of the velocity of a particle moving under the influence of a central force.

The obvious way of approaching the question about the rates of change of v_r and v_θ is to examine first of all the force acting along the radius line. This is a combination of the central force, $F(r)$ and the centrifugal force $mr\omega^2$. Obviously,

$$m\,\mathrm{d}v_r/\mathrm{d}t = F(r) - mr\omega^2,$$

but if the angular velocity were zero the central force would produce an acceleration $F(r)/m = \mathrm{d}^2 r/\mathrm{d}t^2$, so that

$$\mathrm{d}v_r/\mathrm{d}t = \mathrm{d}^2 r/\mathrm{d}t^2 - r(\mathrm{d}\theta/\mathrm{d}t)^2,$$

providing the acceleration for one of the components of the velocity.

In this central force problem there is a force acting in a direction perpendicular to the direction of r (except when the path is circular) and the angular velocity changes as the particle travels around the orbit. We represent this as a force, F_θ, acting in a direction perpendicular to the radius line. This provides a torque Γ ($= rF_\theta$) which gives rise to the angular acceleration. We know that the torque is the rate of change of angular momentum, $mr^2\omega$, so that

$$\Gamma = rF_\theta = m\,\frac{\mathrm{d}}{\mathrm{d}t}(r^2\omega),$$

giving

$$\frac{F_\theta}{m} = \frac{\mathrm{d}v_\theta}{\mathrm{d}t} = \frac{1}{r}\frac{\mathrm{d}}{\mathrm{d}t}\left(r^2\frac{\mathrm{d}\theta}{\mathrm{d}t}\right).$$

12.1.1   Radial, angular, and tangential accelerations

The expressions that we have obtained for the radial and angular velocity and acceleration of a particle or planet describe the components of its actual velocity and acceleration, its tangential velocity, and its acceleration. An alternative derivation of the radial and angular acceleration equations can be made by means of a simple vector argument. Let the situation of the particle at the beginning and end of a time interval δt be defined by the lines OQ and OP in Fig. 12.2. The point O is the situation of the centre of force. The radial and angular velocities at Q are the components of the particle velocity parallel and perpendicular to the radial vector, $v_\parallel$ and $v_\perp$, and those at P are $v_\parallel + \delta v_\parallel$ and $v_\perp + \delta v_\perp$ (we have altered the

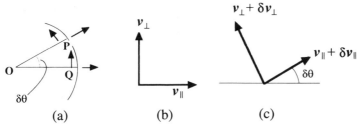

Fig. 12.2 (a) The position and velocities of a particle moving under the influence of a central field of force; (b) the velocity vector diagram for the particle when it is at the position Q; (c) the velocity vector diagram for the position P.

subscript θ to $\perp$ and the subscript r to $\parallel$ to emphasize the nature of the component velocities). The change in angular position between the points Q and P is $\delta\theta$, as shown in Fig. 12.2(a). The velocity vector diagrams for the particle at the positions Q and P are given in Figs. 12.2(b) and (c) respectively. The difference between the horizontal components of Figs. 12.2(b) and (c) represents the increase in the velocity parallel to the radius lines. Because the orientations of $v_{\parallel}$ and of $v_{\perp}$ have changed, this change in velocity involves $v_{\perp}$ as well as $v_{\parallel}$. The change in velocity parallel to the radius line in the time interval δt is thus

$$\delta v_r = (v_{\parallel} + \delta v_{\parallel})\cos\delta\theta - (v_{\perp} + \delta v_{\perp})\sin\delta\theta - v_{\parallel}.$$

As $\delta\theta$ is small $\cos\delta\theta = 1$ and $\sin\delta\theta = \delta\theta$ so that

$$\delta v_r = \delta v_{\parallel} - v_{\perp}\,\delta\theta + \delta v_{\perp}\,\delta\theta.$$

Neglecting the last term because it's the product of two very small quantities, we have, dividing by δt and taking the limit as $\delta t \to 0$,

$$\frac{dv_r}{dt} = \frac{dv_{\parallel}}{dt} - v_{\perp}\frac{d\theta}{dt} = \frac{d^2 r}{dt^2} - r\left(\frac{d\theta}{dt}\right)^2,$$

since $v_{\parallel} = dr/dt$ and $v_{\perp} = r\omega$. Resolving the velocities vertically leads to an acceleration perpendicular to OQ in the form

$$\frac{dv_{\theta}}{dt} = \frac{1}{r}\frac{d}{dt}\left(r^2\frac{d\theta}{dt}\right).$$

These two equations of motion, for the components of the tangential acceleration parallel and perpendicular to the radial vector, form the basis of mathematical exercises which show, for example, that the closed orbits allowed when the particle moves in a central field of force are elliptical or, as a limiting case, circular. The square of the tangential velocity is obtained by summing the squares of its components parallel and perpendicular to the radial vector and, in the same way, the sum of the squares of the acceleration components gives the square of the tangential acceleration. This second derivation of the equations reminds us that vectors are useful quantities.

The concept of angular momentum as a property of motion in two dimensions is particularly simple. The angular momentum is represented by a vector which is directed perpendicular to the plane in which the motion takes place. As long as the forces and linear impulses which act on the system have their directions restricted to the plane of motion, their effect is to produce changes in the magnitude of the angular momentum vector, L, without altering its direction. The spatial orientation of L remains constant with respect to the plane of motion. If we describe the plane of motion as the xy-plane then L is always directed parallel to (or along) the z-axis. On the other hand, we have seen that a spinning body travelling in a closed orbit has a total angular momentum, the combination of its orbital angular momentum and its spin angular momentum, which in general isn't perpendicular to the plane of the orbit. Only when the body's spin was about an axis perpendicular to the plane of the orbit would the vector representing total angular momentum lie parallel to the z-axis. When the axis of spin is oriented in any other direction the total angular momentum won't be parallel to the z-axis; L_T will have x- and y-components and may rotate about the z-axis in a precessional motion. L_T will have x- and y-components because the spin angular momentum has x- and y-components which cause L_T to tilt away from the direction of the z-axis.

The spin of a body in orbital motion provides an exception to the 'rule' that we have used so far, that L is perpendicular to the plane of the orbit. This exception arises because the spin (except in two special cases) is about an axis which isn't parallel to the z-axis; if we want to describe the spin angular momentum in the coordinate system the xy-plane of which contains the orbit, we must introduce x- and y-components of the angular momentum. This observation provides us with the idea that if L has x- and y-components the simple two-dimensional picture we have of orbital angular momentum very probably isn't appropriate in three-dimensional problems. Many of the examples that we have dealt with appear to have been three-dimensional, but they were of a form that made the methods developed for two-dimensional problems quite adequate to provide solutions. What approach should we take if the problem can't be simplified in this way? The answer lies in the application of simple mechanical principles.

12.2 Angular momentum and impulse relations

Example 65. A point particle of mass m, attached to a pivot by a light rod of length r, is travelling with a velocity v in a circular orbit in the xy-plane. The particle is subjected to an impulse J, the direction of which lies in the xy-plane. What is the form of the subsequent motion? What is the change in angular momentum of the particle?

Initially, the angular momentum of the system (in scalar terms) is mvr, or as a vector

$$L = r \times p,$$

where $p = mv$ is the linear momentum of the particle. This is illustrated in Fig. 12.3, which shows the vectors v, J, and L. The increase in linear momentum produced by the impulse will be

$$\delta p = J$$

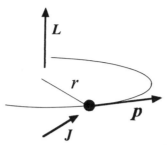

Fig. 12.3 A particle travelling in an orbit in the x–y plane with a velocity v subjected to an impulse J which is confined to the plane of the orbit.

so that the change in velocity will be $v = J/m$. If J is in the same direction as v, the new velocity will be $v_1 = v + J/m$. This increase in velocity will cause the particle to move faster in its orbit if $|J| > 0$, and slower if $|J| < 0$.

When the impulse J (in the plane of the orbit) has a component perpendicular to the direction of v—in other words, it has a radial component—that component has no effect on the motion because of the rigidity of the rod defining the path radius. We conclude that when such an impulse has components $J_\parallel$ and $J_\perp$ parallel and perpendicular to v, $J_\parallel$ increases the particle velocity by $J_\parallel/m$ and $J_\perp$ has no effect. The change in the angular momentum as a result of the impulse is the angular impulse, $\delta L = r \times J$, and is parallel to the z-axis. The angular momentum has its magnitude altered, but its orientation, perpendicular to the plane of the orbit, isn't altered. Once again, a superficially three-dimensional problem reduces to a two-dimensional one.

The situation would be a little different if the particle were not rigidly attached to the pivot by a rod. The impulse would force it out of its initial stable orbit (which was an 'equilibrium state') and, if the equations of motion allowed, it would progress to another state of equilibrium in a stable orbit.

Example 66. A particle of mass m travels along a straight line in the xy-plane with a velocity v. Its distance of closest approach to an origin O is R. What is its angular momentum measured with respect to the point O? If the particle is subjected to an impulse J which is in the z-direction, what is the subsequent motion of the particle, and how is its angular momentum changed?

In the initial state the particle has a linear momentum $p_1 = mv$ and p_1 lies in the xy-plane, collinear with v. The initial angular momentum of the particle with respect to O then has a magnitude of $mvR = m\omega R^2$, where we have introduced an angular velocity ω ($= v/R$). The impulse J is in the z-direction, so that the change in momentum it produces is also in the z-direction, $|J| = \Delta p_z$, or since p_z is initially zero, $J = p_z$. The new linear momentum of the particle is $p = |p| = (p_1^2 + p_z^2)^{1/2}$, and is directed along a line inclined at an angle $\theta = \tan^{-1}(p_z/p_1)$ out of the xy-plane. The new angular momentum of the particle is pR.

The initial angular momentum L_1 is a scalar representation of the vector product $m\,(R \times \omega \times R) = L_1$, a vector which is directed perpendicular to the *xy*-plane. The impulse J in the *z*-direction produces an angular impulse $J_\omega = R \times J$ which is directed parallel to the *xy*-plane. Since the angular impulse is equal to the change in angular momentum, the change in angular momentum is $L_2 = R \times J$, directed parallel to the *xy*-plane. L_1 and L_2 are directed at right angles to one another, so that the resulting angular momentum is of magnitude $|L| = |L_1 + L_2|$ and is directed at an angle of $\tan^{-1}(L_1/L_2)$ to the *xy*-plane, or at an angle of $\tan^{-1}(L_2/L_1)$ to the direction of L_1.

The effect of the angular impulse is to alter the magnitude of the angular momentum vector of the system *and* to change its direction. The particle's path is tilted out of the *xy*-plane.

Example 67. A point particle of mass m, attached to a pivot by a light rod of length r, travels with speed v ($= |v|$) in a closed orbit in the *xy*-plane. The particle is subjected to an impulse J which is directed perpendicular to the *xy*-plane. What is the form of the subsequent motion?

This example is closely similar to Example 66, the only difference being that the particle is attached to the pivot and moves in a circular path in the *xy*-plane. Again, the impulse has no *x*- or *y*-components, only a *z*-component. The impulse produces a *z*-component of the particle's momentum

$$p_z = mv_z = |J|$$

directed perpendicular to the plane of the orbit, as shown in Fig. 12.4. The momentum in the *xy*-plane, p_1, is unaffected by the impulse. The momentum of the particle after the impulse is

$$|p_f| = (p_1^2 + p_z^2)^{1/2}$$
$$= (p_1^2 + J^2)^{1/2}$$

and is directed at an angle of $\theta = \tan^{-1}(p_z/p_1)$ to the *xy*-plane. This means that the particle is travelling in a direction inclined at an angle θ to the *xy*-plane. In terms of the change in the angular momentum vector, $\delta L = r \times J$, and examination of Fig. 12.4 shows that δL is directed in the negative *x*-direction. The angular momentum after the impulse has taken effect is

$$L_f = L + \delta L = L + (r \times J)$$

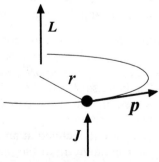

Fig. 12.4 An impulse J applied in the *z*-direction to a particle moving in an orbit confined to the *xy*-plane.

and L_f is inclined at an angle θ with respect to the z-axis. Since the particle is travelling in a closed orbit and L_f is perpendicular to the new plane in which the orbit lies, this indicates that the effect of the angular impulse is to tilt the plane of the orbit. This change can be explained in terms of the velocity of the particle after the impulse. When a body travels in a closed orbit its velocity vector must lie always in a plane. If the velocity vector is tilted out of the plane which contains the original orbit, it must lie within a new plane of motion. The original orbit tilts so that it lies within this new plane.

We conclude that the orientation, as well as the magnitude, of the angular momentum vector is altered by an angular impulse with a non-zero component perpendicular to the plane of the orbit. The effect observed is that the orbit tilts so that it lies in a plane different from that in which it lay originally. The particle continues in its orbital path because it is connected to the pivot by the rod. The angular momentum vector is still perpendicular to the plane of the orbit, but has changed its direction because the plane of the orbit has been altered.

Example 68. The governor of a steam engine regulates the speed at which it runs. In essence, it consists of a vertical rod, the upper end of which is hinged to the top of a shorter light rod of length r, the other extremity of which carries a mass m. The engine drives the vertical rod in rotation about its length with an angular velocity ω so that the hinged rod moves away from the vertical, the two rods make an angle θ with one another, and the mass at the hinged rod end travels around a closed circular path of radius $r \sin \theta$. If the mass is 1 kg, the rate of rotation is 60 rpm, and r is 30 cm, what is the tension in the hinged rod? What is the angular momentum of the system?

The forces acting on the governor are the weight of the mass, the tension in the light rod, and the centrifugal force. These are shown schematically in Fig. 12.5. If the angular displacement of the rod away from the vertical is θ, we may resolve these forces to give

$$T \sin \theta = \omega^2 mr \sin \theta$$

and

$$T \cos \theta = mg,$$

so that

$$T = \omega^2 mr$$

and

$$\cos \theta = g/r\omega^2.$$

The tension in the rod is thus $4\pi^2 \times 1 \times 0.3 = 11.8$ N. The angle that the rod makes with the vertical is 34.16°.

To find the angular momentum of the system we need to re-examine it in terms of the vectors involved; say, at the time when the position of the system is as shown in Fig. 12.5. The linear velocity of the particle is given by $v = \omega \times r$ and is directed perpendicularly into or out of the plane of the diagram in Fig. 12.5, depending on

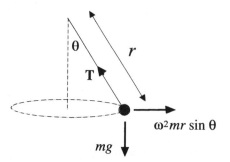

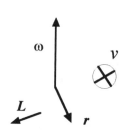

Fig. 12.5 A governor represented as a conical pendulum with length r carrying a mass m which rotates with angular velocity ω in a circular path of radius $r \sin \theta$.

Fig. 12.6 The vectors which define the angular momentum L of the governor.

the sense of ω. The corresponding vector diagram is given in Fig. 12.6. The angular momentum of the system is given by

$$L = m(r \times v)$$

or

$$L = mr \times (\omega \times r).$$

Since the product $\omega \times r$ has a magnitude of $\omega r \sin \theta$ and the angle between v and r is 90°, the magnitude of L is $mr^2\omega \sin \theta$. This has a value of 0.32 N s m. From Fig. 12.6 it is obvious that L is not parallel to the direction of ω. The angular momentum vector is no longer directed perpendicular to the plane in which the circular motion takes place. What happens, then, at some later time when the mass has moved out of the plane of the diagram in Fig. 12.5? By going through another vector multiplication, or by an appeal to symmetry, we see that the effect of rotation of the mass is to rotate the angular momentum vector, L. This rotation of L about an axis which is not its own direction is called precession. Obviously, L precesses with angular velocity ω about the vertical axis of the system. Although the magnitude of L is constant, L itself is no longer constant because its orientation is changing continuously as it rotates. If L is variable—that is, L changes with time—then $dL/dt \neq 0$. A torque must be acting on the system to maintain the notion, and that torque is defined by

$$\Gamma = dL/dt.$$

We can find the value of the torque quite easily by drawing a vector diagram showing the precession of L (Fig. 12.7(a)), then resolving L parallel and perpendicular to ω; that is, into L_y parallel to ω and L_x perpendicular to ω (as in Fig. 12.7(b)). L_y is constant in magnitude and direction, but L_x, although constant in magnitude, varies in orientation with time. In a time interval δt, L_x rotates through an angle $\omega \delta t$, as shown in Fig. 12.7(c). The difference between $L_x(t)$ and $L_x(t + \delta t)$, δL_x, is the third side of the triangle shown in Fig. 12.7(c). From this diagram, $\delta L_x = 2L_x \sin(\frac{1}{2} \omega \delta t)$, and as $\delta t \to 0$,

$$dL_x/dt = \omega L_x = \omega L \cos \theta = mr^2\omega^2 \sin \theta \cos \theta.$$

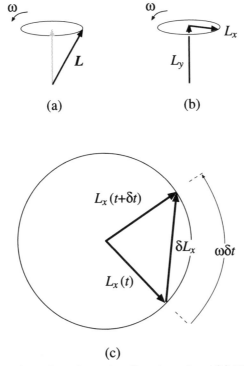

(a)   (b)

(c)

Fig. 12.7 (a) The precession of **L** about the direction of **ω**. (b) The components of **L** parallel and perpendicular to **ω**. (c) The vector diagram for the calculation of d**L**/d*t*.

Because the angle between **ω** and **L** is $\frac{1}{2}\pi - \theta$, this equation implies that we can write

$$\mathrm{d}L_x/\mathrm{d}t = \mathrm{d}L/\mathrm{d}t = \boldsymbol{\omega} \times \boldsymbol{L} = \boldsymbol{\Gamma}$$

where **Γ** is the torque required to maintain the motion.

   Finally, we may say that the magnitude of the angular momentum is 0.32 J s, it is directed at an angle of 55.84° with respect to the direction of **ω**, and it precesses about the direction of **ω** (in this case the vertical direction) at a rate of 2π s^{-1}, or 60 rpm.

12.3   Some centrifugal forces produce torque!

We could have made a much more simple-minded approach to find the torque in the governor system if we hadn't been so interested in the angular momentum. If we redraw Fig. 12.5, showing the path of the governor weight (see Fig. 12.8), it becomes obvious that the rotation of the weight with angular velocity **ω** gives rise to a centrifugal force $F = m\omega^2 r \sin\theta$, which is directed perpendicular to the axis of

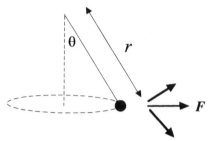

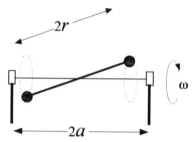

Fig. 12.8 The centrifugal force arising from the rotation of the governor weight.

Fig. 12.9 A shaft of length $2a$, carrying a dumb-bell assembly of length $2r$ and mass $2m$, rotating with angular velocity ω.

rotation. The torque acting at the end of the rod, $r \times F$, has a magnitude of

$$\Gamma = Fr \cos \theta,$$

so that

$$\Gamma = m\omega^2 r^2 \cos \theta \sin \theta,$$

which is exactly the same result as we obtained in Example 68. Because this torque is so simply related to the centrifugal force it might be called the centrifugal torque, but that's hardly allowed; centrifugal implies movement away from a centre, while torque is associated with movement around a centre, so that the term 'centrifugal torque' seems self-contradictory.

Example 69. Two point masses, each of mass 100 g, are joined by a light rod of length 60 cm. This assembly is mounted at the centre of a shaft of length 1 m which passes through the centre of mass of the assembly. The shaft is carried by a bearing at each of its ends and rotates at 300 rpm. If the angle between the shaft and the rod carrying the masses is 30°, what force acts at the bearings as a consequence of the rotational motion?

The rotating system is illustrated in Fig. 12.9, in which the dimensions and angular velocity are represented as shown by m, r, a, and θ, and the directions of a set of Cartesian axes, the origin of which is at the centre of the shaft, are shown. By considering one end only of the assembly, we may say that it produces a torque of $mr^2 \omega^2 \cos \theta \sin \theta$ at the centre of the shaft. Because of the symmetry of the system, it's obvious that the two weights produce a torque of $2mr^2\omega^2 \sin \theta \cos \theta$ at the shaft centre. This torque must be compensated by a torque produced by forces at the bearings. If these forces are represented by F, then

$$2Fa = 2mr^2\omega^2 \cos \theta \sin \theta$$

or

$$F = [(mr^2\omega^2)/a]\cos \theta \sin \theta$$

a dynamic bearing force which has to be added to the static bearing force to find the total force acting at the bearing at a particular time. For the system described in the example, $m = 0.1$, $r = 0.3$, $a = 0.5$, and $\theta = 30°$, so that $F = 19.5$ N. This is about 20 times greater that the static force, $\frac{1}{2}(2mg)$, which acts on each bearing. If θ were reduced to 5°, this dynamic force would be about four times the static force.

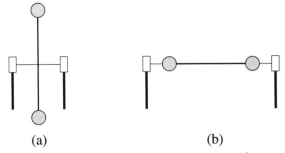

Fig. 12.10   The dumb-bell–shaft rotating system: (a) with $\theta = \frac{1}{2}\pi$, and (b) with $\theta = 0$.

12.4   Principal axes

In Example 69, the form of the expression for F, involving the term $\sin \theta \sin \theta$, indicates that the bearing force and the centrifugal force have their maximum values for $\theta = 45°$. Furthermore, the minimum value of F, which corresponds to $\theta = 0$ or $\frac{1}{2}\pi$, is zero. These two situations are shown in Fig. 12.10, whence it is obvious that for $\theta = 0$ the centrifugal force is zero, so that it can produce no torque. When $\theta = \frac{1}{2}\pi$ the centrifugal forces acting at either end of the dumb-bell are equal and opposite, with the result that the torque due to them is zero. When the torque is zero, the system is described as dynamically balanced.

Inspection of Fig. 12.10 shows clearly that zero torque due to centrifugal force corresponds to particular axes of rotation which are perpendicular to each other. This leads to the idea that we might define a set of Cartesian axes, with the origin at the centre of mass of a system such that rotation about any one of the axes will produce no torque due to centrifugal forces. We call these axes the principal axes of the system. For simply shaped, homogeneous bodies, the principal axes will be axes of symmetry. These may be selected by observation, as illustrated for a rectangular rod and for a circular disc in Fig. 12.11. When such bodies are rotated about their principal axes the lines of action of the centrifugal forces all pass

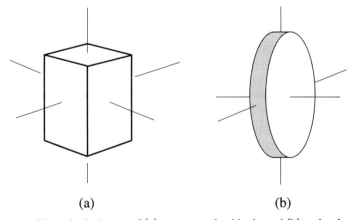

(a)   (b)

Fig. 12.11   The principal axes of (a) a rectangular block, and (b) a circular disc.

through the centre of rotation; they produce no torque and the system can be described as 'balanced'.

12.4.1 Products of inertia

There is another way of describing the torque due to centrifugal force that is of interest. If we return to Fig. 12.9, we see that in the situation shown in the diagram the x- and y-coordinates of one of the point masses are $r \cos \theta$ and $r \sin \theta$ respectively. We could have written the torque in the form

$$\Gamma = 2mr^2\omega^2 \sin \theta \cos \theta = 2\omega^2 mxy.$$

We can generalize from this observation because we could say that the dumb-bell with masses m at each end is a representation of two mass elements δm lying in the plane of a lamina. If that were so, then to find the torque we would have to sum the contributions from every mass element making up the lamina; that is,

$$\Gamma = \omega^2 \int xy \, dm.$$

This torque is directed along the z-axis. If the rotating body is two-dimensional, there is a further torque component directed along the y-axis (for rotation about the x-axis). We have then the torque components Γ_z and Γ_y, defined as

$$\Gamma_z = \omega^2 \int xy \, dm \quad \text{and} \quad \Gamma_y = \omega^2 \int xz \, dm.$$

The integrals $\int xy \, dm$ and $\int xz \, dm$ are known as the products of inertia of the body (there is, of course, a third product of inertia, $\int yz \, dm$, which we haven't considered because we chose the rotating body as a lamina). We now have a new definition of the principal axes; the principal axes of a body are those axes for which the products of inertia are zero. Rotation about one of the principal axes doesn't produce any torque. Alternatively, a system is dynamically balanced when its products of inertia about the axis of rotation are zero.

12.5 Angular velocity components

This definition of the principal axes of a body makes it possible to view the angular momentum of the rotating dumb-bell in another way. It's quite clear that the principal axes of the dumb-bell lie along its length and perpendicular to its length, all passing, of course, through its centre of mass. The angular velocity ω is directed at an angle θ to the length of the dumb-bell; that is, at an angle θ to one of the principal axes. There is no reason why we should not describe the system in terms of a coordinate system which coincides with the principal axes of the dumb-bell rather than the coordinate system which was defined by the axis of rotation. Let's say that the length of the dumb-bell is the x-axis, and that the y- and z-axes are perpendicular to the length, as shown in Fig. 12.12. When we choose this system of

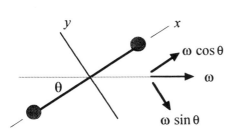

Fig. 12.12 The rotating dumb-bell described in terms of a coordinate system which coincides with the principal axes of the dumb-bell.

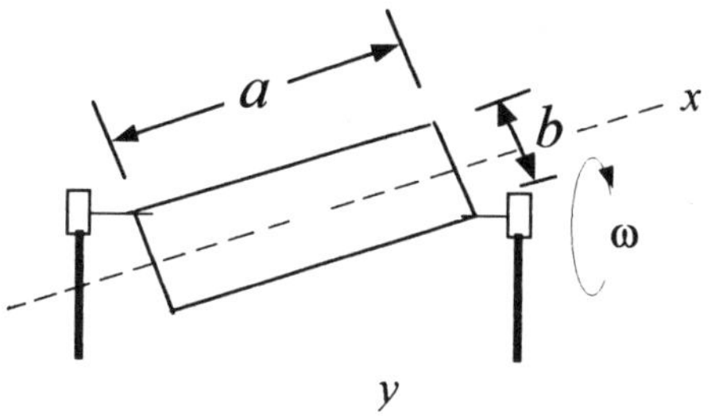

Fig. 12.13 A rectangular plate being rotated with angular velocity ω about a diagonal.

coordinates, we can represent ω by its components, $\omega_x = \omega \cos \theta$ and $\omega_y = \omega \sin \theta$, along and perpendicular to its length respectively.

In Example 68 we derived the expression for the angular momentum of a 'single-ended dumb-bell' as $L_s = mr^2\omega \sin \theta$, and showed that the angular momentum vector was directed at right angles to the length of the dumb-bell. The only difference between that case and the one of the full dumb-bell is that there is an angular momentum contribution of $mr^2\omega \sin \theta$ from each end of the dumb-bell, so that $L = 2L_s = 2mr^2\omega \sin \theta$ for the system shown in Fig. 12.12. The direction of the angular momentum vector is still perpendicular to the length of the dumb-bell, as shown in Fig. 12.12. This angular momentum could be written as $2mr^2\omega_y$ and, since $2mr^2$ is the moment of inertia of the dumb-bell about the y-axis, we may equate the angular momentum L with $I_y\,\omega_y$, where I_y is the moment of inertia of the dumb-bell about the y-axis (which is one of the principal axes). Once we have chosen the principal axes of the dumb-bell, the y- and z-axes rotate about the length of the dumb-bell with angular velocity ω_y, just as does the angular momentum vector. This leads us to identify L with the y-component of the angular momentum, so that we may write

$$L = L_y = I_y\,\omega_y$$

This identity is justified when we examine the other components of the angular momentum. The x-component is zero because the moment of inertia of the point mass – light rod system along its length is zero, so that $L_x = I_x\,\omega_x = 0$, even though $\omega_x = \omega \cos \theta \neq 0$. There is no z-component of the angular velocity, $\omega_z = 0$, because we have chosen the perpendicular component of ω to be ω_y, so that $L_z = I_z\,\omega_z = 0$. The angular momentum vector is directed along the y-axis, so that the identity of L with L_y is complete. It should be borne in mind that the dumb-bell is being driven in rotation with the angular velocity ω about an axis inclined at an angle θ with respect to the x-axis, so that the y-axis is being rotated with angular velocity ω; that is, L (which has been identified with L_y) precesses about the direction of ω with a frequency of $\omega/2\pi$.

Example 70. A uniform rectangular plate has a mass of 1 kg and sides of 20 and

10 cm. It is mounted on a shaft (a light rod) which rotates at 300 rpm. If the axis of the shaft coincides with one of the diagonals of the rectangle, what is the angular momentum of the plate as it rotates?

It's obvious, because of the symmetry of the plate, that two of its principal axes are the lines that bisect its sides. These are shown in Fig. 12.13 as dashed lines, labelled as the x- and y-axes. If the sides of the rectangle were a and b, the moments of inertia about these principal axes would be $I_y = \frac{1}{12}ma^2$ and $I_x = \frac{1}{12}mb^2$, where m is the mass of the rectangle. The angle between the axis of rotation and the y-axis is $\theta = \tan^{-1}(a/b) = \tan^{-1}(2) = 63.435°$. Since $\tan\theta = \sin\theta/\cos\theta$, we can say that $\sin\theta = 2\cos\theta$ and the y- and x-components of the angular velocity, ω_y and ω_x, are related by

$$\omega_y = \omega\cos 63.4° = \tfrac{1}{2}(\omega\sin 63.4°) = \tfrac{1}{2}\omega_x.$$

For the plate in question $(b/a) = 2$ so that $I_y = 4I_x$. The components of the angular momentum are then $L_y = I_y\omega_y = 2I_x\omega_x = 2L_x = 4.683 \times 10^{-2}$ N s m. The angular momentum of the plate can then be described in terms of a vector L the magnitude of which is the square root of the sum of L_y^2 and L_x^2 and the direction of which relative to the x-axis is given by the angle $\phi = \tan^{-1}(L_y/L_x)$: L therefore has a magnitude of 5.236×10^{-2} N s m, and is directed at an angle of $63.435°$ with respect to the x-axis. Its orientation with respect to the axis of rotation is $63.435° - (\tfrac{1}{2}\pi - 63.435°) = 36.870°$, and it precesses about the axis of rotation with an angular velocity of 10π s^{-1}.

This complete specification of the angular momentum of the rotating plate could be used to find out the torque produced by the rotation. We have seen that $\Gamma = \omega \times L$, so that the magnitude of the torque is simply $\Gamma = \omega L \sin\phi$, where ϕ is the angle of precession of the angular momentum vector about the axis of rotation. Here we have $\phi = 36.87°$, $\omega = 10\pi$, and $L = 5.236 \times 10^{-2}$, leading to $\Gamma = 0.9869$ N m, directed perpendicular to the plane of the plate.

There is an alternative way of finding the torque which doesn't involve any specification of the angular momentum. We may, if we wish, take the expression for Γ in the form of its components so that, for example,

$$\Gamma_z = \omega_x L_y - \omega_y L_x$$

which, since $L_y = I_y\omega_y$ and so on, may be written as

$$\Gamma_z = \omega_y\omega_x(I_y - I_x) = \omega^2\cos\theta\sin\theta(I_y - I_x).$$

In this example $\omega = 10\pi$, $\theta = \tan^{-1}(2)$, and $(I_y - I_x) = 2.5 \times 10^{-3}$, so that $\Gamma_z = 0.9869$ N m along the z-direction (perpendicular to the plane of the plate). In this way the torques due to centrifugal force may be found quickly without the necessity of calculating the magnitude and direction of the angular momentum, a much simpler process.

As a corollary to this last observation, the form of the equation defining Γ_z tells us that the torque Γ_z due to centrifugal force will be zero provided that $I_y = I_x$. Thus, for example, there will be no torque when the body rotating about an axis (but not necessarily what might be thought to be a principal axis) passing through its centre of mass has a high degree of symmetry which makes $I_y = I_x$. Amongst laminas it's easy to see that the square and the circle produce zero torque in that situation, two cases in which calculation bears out an observation based in common sense. For the circle any diameter is a principal axis, but we would generally think of the principal axes of a square lamina as the lines which are the perpendicular bisectors of its sides. The other components of Γ, of interest in the case of three-dimensional bodies, are zero when $I_z = I_x$ and when $I_z = I_y$, respectively. The requirement for zero torque, that $I_x = I_y = I_z$, means that amongst homogeneous solid bodies, only the sphere produces no centrifugal torque when it is rotated about any axis (which hasn't been chosen as a principal axis) passing through its centre of mass. Alternatively, for a homogeneous sphere any axis of rotation passing through its centre of mass (any diameter) may be taken as one of its principal axes.

12.6 Kinetic energy of rotation

Example 71. A uniform rectangular plate has a mass of 1 kg and sides of 20 and 10 cm. It is mounted on a shaft (a light rod) which rotates at 300 rpm. If the axis of the shaft coincides with one of the diagonals of the rectangle, what is the (kinetic) energy of the plate as it rotates?

So far, we have dealt with the kinetic energy of rotation by equating it with $\frac{1}{2}I\omega^2$, since there has been no ambiguity about what is meant by I. That's because we have chosen the axes of rotation to coincide with, or to be parallel to, a principal axis of the body under consideration. In other words, we have calculated, say, I_x, and chosen the axis of rotation as the x-axis so that our expression for the kinetic energy was implicitly $\frac{1}{2}I_x\omega_x^2$ $(= L_x^2/2I_x)$. Similarly, had we taken the y-axis to be the axis of rotation, we would have obtained the expression $\frac{1}{2}I_y\omega_y^2$ for the kinetic energy of rotation. If the axis of rotation were not one of the principal axes, then we would have the choice of working out the moment of inertia about the rotation axis, involving the complication of calculating the products of inertia, or of taking the components of ω along the principal axes and multiplying their squares by the (relatively simply calculated) moments of inertia about the corresponding principal axis. In general, this second approach is easier than the first, and so we write the expression for the kinetic energy of rotation as

$$T = \tfrac{1}{2}\left(I_x\,\omega_x^2 + I_y\,\omega_y^2 + I_z\,\omega_z^2\right),$$

where the I's are the principal axis moments of inertia and the ω's are the components of ω along the principal axes.

In this case we have $I_y = \frac{1}{12}mb^2 = 3.333 \times 10^{-3}$ kg m^2, $I_x = \frac{1}{12}ma^2 = 8.33 \times 10^{-4}$ kg m^2, $\omega_y = \omega \cos\theta = 10\pi\cos 63.435° = 14.050$ s^{-1}, and $\omega_y = \omega\sin\theta = 28.099$ s^{-1}. Substitution into the kinetic energy equation gives

$$T = \tfrac{1}{2}\left(I_x\,\omega_x^2 + I_y\,\omega_y^2\right) = \tfrac{1}{2}(0.6579 + 0.6579)$$
$$= 0.6579\,\text{J},$$

since $\omega_z = 0$. It's interesting to notice that the two terms in the expression for the kinetic energy of a rectangle rotated about a diagonal are always equal, irrespective of the value of b/a.

12.7   Free body rotation

In examples 68–71 it has been shown that when a body is rotated about a fixed axis which is a non-principal axis a torque is produced, the origin of which is centrifugal force. This torque has to be cancelled out by a compensating torque produced at the bearings which fix the axis of rotation. The effect of this torque is to cause the angular momentum vector to precess about the direction of the axis of rotation, which is the direction of the angular velocity vector. This phenomenon is obviously of importance when we are dealing with rotating machinery for which the torque due to centrifugal forces will produce energy losses at the bearings. Part of the energy loss will be manifested as wear of the bearings which, with an unbalanced rotor, would deteriorate catastrophically.

Of course, a body can be rotated even though there is no fixed axis of rotation defined by an axle mounted on bearings. In ball games, when the ball is given spin the axis of rotation is chosen to be in a particular direction in the body of the ball. In most games the ball is not smoothly spherical and even in those in which it should be is unlikely to be exactly so. It will have well-defined principal axes which have fixed directions relative to its shape and mass distribution. In order to avoid the complications due to the interaction between the ball and the medium through which it travels (usually air) we assume that it is projected into an evacuated space. Once the ball is released, or breaks contact with whatever impels it, there is no mechanism to produce a torque analogous to the bearing torque. In the absence of a torque the angular momentum is constant. The angular momentum vector doesn't change its magnitude or its direction and could be called a constant of the motion. If the ball rotated about a principal axis, which would be the case every time if the ball were a homogeneous sphere, the angular momentum vector and the angular velocity vector would be collinear. As we have seen from the previous examples, unless the axis of rotation was a principal axis the angular momentum vector and the angular velocity vector wouldn't be collinear and there would be a centrifugal torque. What then happens? Because the angular momentum vector cannot change in the absence of an external torque, and there is no such torque, the angular velocity vector has to precess about the direction of the angular momentum vector. For all bodies except the homogeneous perfect sphere, this precession of the axis of rotation gives rise to a wobbling of the body as it follows the path along which it was projected. Because the rotating body is an isolated system, the centrifugal torque is an uncompensated internal torque, giving rise to the precession of ω about the direction of L. In the absence of dissipative mechanisms (leading to energy losses) the system would continue in this state of whirling motion for ever.

12.8 Rotating frames of reference

Example 72. A megalomaniac dictator, whose domain was on the equator, ordered the construction of the highest building in that part of the world. It was made 300 m tall and straddled the equator. At the topping-out ceremony a bolt fell from the roof of the building. Where did it land on the ground?

The initial reaction to this question is to assume that the bolt had zero velocity both in the horizontal and the vertical directions as it left the roof. Since the only force acting on it (neglecting the resistance to motion due to the air) would be the force of gravitational attraction, it would suffer zero acceleration horizontally and its acceleration would be vertically downwards. These observations lead us to conclude that it would land right alongside the wall of the building directly below the point from which it fell. This elementary analysis would be adequate provided that the tower and the surface of the earth were stationary. Then we recollect that the earth is spinning about a N–S axis and completes one revolution in one day; that is, its angular velocity is $\omega = 2\pi/(24 \times 3600)\,\mathrm{s}^{-1}$. This rotational motion of the earth means that its surface, and constructions fixed to its surface, will have an angular momentum. The bolt, with mass m, will have an angular momentum on the surface of the earth at the equation of $mR^2\omega$, where R is the earth's radius, and an angular momentum of $m(R + 300)^2$ at the top of the tower. During its fall the bolt has lost angular momentum. We can quantify this loss of angular momentum because at a distance r from the earth's axis of rotation the angular momentum of a body of mass m is

$$L = mr^2\omega \qquad \text{and} \qquad \frac{\mathrm{d}L}{\mathrm{d}t} = m\omega 2r \frac{\mathrm{d}r}{\mathrm{d}t} = \Gamma$$

where Γ is the torque required to produce such a rate of change of L. The velocity of the body in a direction towards the axis of rotation is $\mathrm{d}r/\mathrm{d}t$, so that we may write $v_r = \mathrm{d}r/\mathrm{d}t$ and $\Gamma = 2mr\omega v_r$. A torque can be represented as the product of a tangential force, F, and the distance from the axis of rotation at which it is applied. In this case the distance is simply r, so that $\Gamma = rF$ and

$$F = 2m\omega v_r.$$

The change in angular momentum which occurs as the body falls has given rise to a force which is perpendicular to the direction of r and directed along the equator. This force produces an acceleration (directed along the equator) of a_t, where

$$a_t = F/m = 2\omega v_r;$$

v_r is the vertical component of the velocity of the body and is simply the product of the time of fall and g, the acceleration due to gravity, $v_r = gt$, with the result that

$$a_t = 2\omega g t = \mathrm{d}v_t/\mathrm{d}t,$$

where v_t is the horizontal (or tangential) component of the body's velocity. If we integrate this equation twice with respect to time, we obtain

$$v_t = 2\omega g \int_0^t t\,\mathrm{d}t = \omega g t^2 \qquad \text{and} \qquad x = \int_0^t v_t\,\mathrm{d}t = \tfrac{1}{3}\omega g t^3,$$

where x is the horizontal displacement of the body from a vertical line of fall. If the height from which the body falls is h, the time of fall is obtained from $h = \frac{1}{2}gt_f^2$ as $t_f = (2h/g)^{1/2}$.

Since the bolt falls $300\,$m, $t_f = 7.825\,$s, and

$$x = \tfrac{1}{3}\omega g (7.825)^3 = 1.138 \times 10^{-1}\ \text{m}.$$

The horizontal displacement of the point of impact of the bolt from the vertical line of fall is thus $11.4\,$cm. The rotation of the earth is towards the east, so that the horizontal displacement will be in the direction of east.

Once again, we see the requirement that angular momentum is conserved gives rise to a torque which has to be compensated by an opposing torque or produces to an observable effect. In this case the torque has the effect of deflecting the path of the bolt from the vertical. In this case this torque has the effect of deflecting the path of the bolt from the vertical. We can think of this deflection as the result of a force, $2m\omega v_r$, which is called the Coriolis force. The angular momentum of the bolt owes solely to the rotation of the earth, so that the force could be said to be the consequence of making observations relative to our position on the surface of the earth. The point on which we stand rotates as the earth rotates. We can imagine, quite plausibly, that our measurements were made in a coordinate system with the earth's centre as its origin (remember that $r = 0$ at the earth's centre). If the axes of the coordinate system were fixed with respect to our position on the earth's surface, it would rotate with the same angular velocity as the earth. As an example we might say that the z-axis of the system passes through the poles, the x-axis through the equator at longitudes of 0 and 180°, and the y-axis through the equator at longitudes of 90°W and 90°E. This coordinate system would provide a framework within which we make our measurements and would be called our frame of reference. It's obvious that this frame of reference is a rotating frame of reference. One of the consequences of the rotation of the frame is the Coriolis force; if our frame of reference were static there would be no Coriolis force and the bolt would fall vertically to the ground.

In choosing the equator as the latitude in Example 72 and a freely falling bolt as its subject, we have made a simplification which may not be obvious. At the equator the force of gravitational attraction is directed at right angles to the axis of rotation of the earth. This leads to v_r being directed perpendicular to the axis of rotation so that the Coriolis force is directed due east. While this is appropriate for a body in free fall, in situations in which the body has a velocity component parallel to the earth's surface there are north or south components of the Coriolis force as well as the easterly one derived in the calculation.

12.9   Coriolis and centrifugal forces

An expression for the Coriolis force may be obtained from a straightforward vector model which involves a rotating vector. We have seen, when dealing with angular momentum, that the rate of change of a precessing angular momentum vector, L, can be expressed simply as $dL/dt = \omega \times L$, where ω is the angular velocity of precession. The same argument used to derive that result when applied to a general vector, say B, would lead to the same form of equation, $dB/dt = \omega \times B$. This, we may say, is the result of our being in a static frame of reference when we observe B; that is, we stand still and watch the vector precessing. The question we have to answer is how our observations would alter if, instead, we were travelling in a frame of reference in which B is static; that is, in a frame of reference which is rotating with an angular velocity ω. The first, and obvious, response is that we would see B as a constant vector. What would happen, though, if B were changing with time? In the rotating frame we would see the rate of change of B as d^*B/dt, where d^*B/dt represents the time derivative of B in the rotating frame. When we step back into the static frame of reference we see a rate of change of B which is the sum of the vectors d^*B/dt and $\omega \times B$, so that

$$\frac{dB}{dt} = \frac{d^*B}{dt} + \omega \times B,$$

where dB/dt represents the rate of change of B observed in the static frame. Since we are interested in finding out about the relationship between forces we can differentiate dB/dt with respect to time to find out how an acceleration in the static frame is related to the corresponding one in the rotating frame. Multiplying these accelerations by the mass of the body involved then gives the relationship between forces in the two different frames. Using this procedure,

$$\frac{d}{dt}\left(\frac{dB}{dt}\right) = \frac{d}{dt}\left(\frac{d^*B}{dt}\right) + \frac{d\omega}{dt} \times B + \omega \times \frac{dB}{dt}.$$

Since d^*B/dt is a vector within the rotating frame, we may write its time derivative as

$$\frac{d}{dt}\left(\frac{d^*B}{dt}\right) = \frac{d^{*2}B}{dt^2} + \omega \times \frac{d^*B}{dt}$$

and we may substitute $(dB/dt) = d^*B/dt + \omega \times B$, giving

$$\frac{d^2B}{dt^2} = \frac{d^{*2}B}{dt^2} + 2\omega \times \frac{d^*B}{dt} + \omega \times (\omega \times B) + \frac{d\omega}{dt} \times B.$$

Let's restrict ourselves to cases in which ω is constant, that is, in which $d\omega/dt = 0$; the last term in this equation disappears. So far, B has been a quite general precessing vector, but if we identify it with the position vector r then its first derivative with respect to time is the velocity, v, and its second derivative is acceleration, a. Remembering that the derivatives d/dt and d^*/dt refer to the static frame and the rotating frame respectively, we may then write the relationship between the accelerations in the two frames as

$$a = a^* + 2\omega \times v^* + \omega \times (\omega \times r),$$

where the superior asterisk refers to quantities observed in the rotating frame. On multiplying by mass m, to obtain an expression for the relation of the forces in the two frames,

$$F = F^* + 2m\omega \times v^* + m\omega \times (\omega \times r).$$

This equation indicates that the force, F^*, observed in a rotating frame of reference is related to the force, F, in a static frame of reference by

$$F^* = F - 2m\,\omega \times v^* - m\,\omega \times (\omega \times r).$$

The last term in this equation is easy to identify with the centrifugal force, while the last but one is a more general form of the Coriolis force than we have seen before. It's obvious that in Example 72 this vector form of the Coriolis force simplifies to $2m\,\omega v^*$ because ω and v^* are perpendicular to one another. This vector form of the Coriolis force can be used to justify the statements made previously about force components in a north or south direction. Those statements can be made subjective with the observation that under the influence of the Coriolis force the path of a body will be deflected towards the right in the northern hemisphere and towards the left in the southern hemisphere. Because the rate of rotation of the earth is small, the Coriolis force is generally small and its effect can be neglected unless the time of flight of the body is large. Obviously, it is not negligible in long range artillery and rocket ballistics. The Coriolis force is perhaps best known because it has been used to explain the general direction of the winds in weather features such as cyclones and anticyclones; that is, clockwise and anticlockwise, respectively, in the northern hemisphere.

We can write the equation for the Coriolis force in a less general form by identifying the force F with the gravitational force, in which case

$$F^* = mg - 2m\,\omega \times v^* - m\,\omega \times (\omega \times r),$$

where mg is the force that would act on the body if we were making observations in the frame of reference that we have called static. It is the force that we would observe if the earth were not rotating. The two other forces in the equation for F^* appear because we are making observations from our position on the surface of the earth which is rotating; that is, in a moving frame of reference. These two forces are sometimes called fictitious forces because they wouldn't be present in the static frame of reference. Alternatively, the static frame of reference is called an inertial frame of reference, and the forces which occur because observations are made in a moving frame of reference are called inertial forces.

Exercises

12.1.   At the end of his act, a juggler filled the stage with 20 plates, each of radius 15 cm and mass 50 g, spinning—with their planes horizontal—at 90 rpm on the top of 2 m long canes. A boy in a box at the side of the stage relished the idea of one of the plates falling to the floor and breaking. From a height of 4 m above the stage he dropped a small rubber ball of mass 25 g which fell vertically and collided with the edge of the plate nearest to him. The ball rebounded vertically without change in the magnitude of its velocity. What were the angular momenta of the plate and the ball immediately before the impact? How great an angular impulse was imparted to the plate, and to the ball? What was the angular momentum of the plate after the collision?

12.2. During a game of conkers, one of the protagonists rotated his conker, C_1, of mass 50 g, at 150 rpm in a horizontal circle, the radius of 30 cm of which was the length of the conker string. His opponent aimed his conker C_2, also of mass 50 g, so that, with its string slack, it travelled vertically upwards following a path perpendicular to the plane of the circle in the vertical plane tangential to the orbit at the point of collision, and collided with the rotating conker. The velocity of C_2 at the moment of impact was $1.88 \, \mathrm{m\,s^{-1}}$ and after impact its velocity was $0.94 \, \mathrm{m\,s^{-1}}$ in the same vertical plane, but directed at 45° to the vertical so that its component in the original plane of the orbit was opposed to the original velocity of C_1. What was the angular momentum of each conker before they collided? What then was the combined angular momentum of the conkers? What was the angular impulse imparted to C_2, what was the total angular momentum after the collision, and what was the orientation of the new plane of motion of C_1?

12.3. A spinning top is made from a circular disc of mass 100 g and radius 20 cm and a light rod which passes through the centre of the disc perpendicular to its plane. The disc is 12 cm from the pointed end of the rod on which the top spins. If the disc spins at 300 rpm and the rod precesses about a vertical axis to describe a cone of half-angle 30°, what is the rate of precession of the top?

12.4. A lecturer, desperate to convince his students, made up an apparatus to demonstrate precession. He used a light rod of length 1 m as an axle at one end of which was a bearing at the centre of a freely rotating metal disc of mass 2 kg and radius 20 cm. A mass of 1.5 kg was fastened to the other end of the axle. The rod was mounted on a universal joint 40 cm from the disc so that it was free to rotate about the horizontal and the vertical axes. When the disc was set in rotation with an angular velocity ω about its centre the assembly precessed with an angular velocity of $3 \, \mathrm{s^{-1}}$. What was the value of ω?

12.5. The flywheel of a motor may be modelled, in a first approximation, as a circular lamina of radius 30 cm and mass 10 kg mounted on a thin light shaft formed by a rod of length 20 cm passing through its centre of mass perpendicular to its plane. The flywheel is carried by two bearings, one at each end of the shaft, and rotates at 1500 rpm. Because of an error in machining during its manufacture the shaft is $\frac{1}{2}°$ away from the axis perpendicular to the plane of the disc. What is the torque exerted on the bearings when the flywheel rotates?

12.6. A flywheel of radius 25 cm and mass 1 kg rotating about a horizontal axis is mounted on bearings carried in two pillar supports set on a table which rotates about a vertical axis at 60 rpm. If the distance from the flywheel to each bearing is 10 cm, what are the forces acting on the two supports? What power is required to maintain the rotation about the vertical axis?

12.7. Assuming that the earth is a perfect homogeneous sphere, make an estimate of the effect of the centrifugal force due to the rotation of the earth on the value of the acceleration due to gravity, g, at the equator and at a latitude of 52°N.

12.8. The Eiffel Tower is situated (approximately) at latitude 49°N. If a ten franc coin is dropped from its top, 300 m above the ground, where will it hit the ground relative to the point from which it fell?

Solutions

12.1. We can make a start by assuming that the plate can be treated as a uniform disc of radius R and mass M. Choosing a coordinate system as shown in Fig. 12.14, and calling the moment of inertia of the disc I, we can say that the initial angular momentum of the plate relative to the z-axis is

$$L_p = kL_{pz} = I\omega = \tfrac{1}{2}MR^2\omega,$$

directed in the positive z-direction. The angular momentum of the ball, referred to the same set of axes, is

$$|L_b| = L_{bx} = mvR = mR(2gh)^{1/2},$$

where m is the mass of the ball, the velocity of which is given by $v = v_z = -(2gh)^{1/2}$, where h is the height through which it has fallen. The angular impulse required to reverse the velocity of the ball is $(-2mvR)$ and this must be equal and opposite to the angular impulse, J, to which the disc is subjected; that is $|J| = 2mvR = 2mR(2gh)^{1/2}$. Since J is directed along the x-axis we may write $|J| = J_x$ and, with δL_{px} the angular momentum gained by the plate, $J_x = \delta L_{px}$. The angular momentum of the plate after the collision is

$$L_{pl} = \left(L_{pz}^2 + L_{px}^2\right)^{1/2} = \left(\tfrac{1}{4}M^2R^4\omega^2 + 8m^2ghR^2\right)^{1/2}$$

which, on substituting the figures given, leads to $L_{pl} = 2.61 \times 10^{-2}$ kg m² s⁻¹.

As a consequence of the impact, the axis of rotation of the plate will tilt so that the new angular momentum vector L_{pl} will make an angle of $\tan^{-1}(L_{pz}/L_{px})$ with the z-axis; that is, $\tan^{-1}(0.3922) = 21.4°$. The plate might well fall off its support, as desired by the boy.

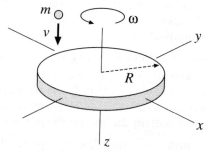

Fig. 12.14 A schematic diagram (pre-impact) of the collision between ball and plate.

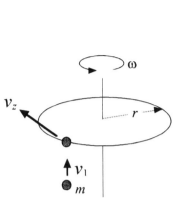

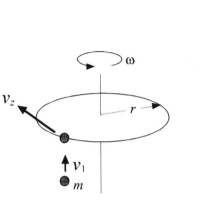

Fig. 12.15 The collision of the two conkers shown schematically.

Fig. 12.16 The spinning top, precessing with its axis of spin inclined at an angle θ to the vertical, with a precession velocity Ω.

12.2. Initially the angular momentum of C_1 is directed vertically in Fig. 12.15 and has a magnitude of

$$L_{11z} = mr^2\omega = 5.66 \times 10^{-2} \text{ J s},$$

whilst that of C_2, directed horizontally, has a magnitude

$$L_{12x} = -mrv_1 = 2.82 \times 10^{-2} \text{ J s}.$$

The total angular momentum of the two conkers is obtained from

$$L_1 = (L_{11z}^2 + L_{12x}^2)^{1/2}$$
$$= mr(v_1^2 + r^2\omega^4)^{1/2} = 5.3 \times 10^{-2} \text{ J s}.$$

After the collision the total angular momentum has the same value, since there is no external torque. Now the angular momentum of C_2 has two components, L_{22x} and L_{22z}, associated with the velocity components v_{2x} and v_{2z}. We have $v_{2x} = -0.94 \sin 45° = -v_{2z}$, so that $L_{22x} = -L_{22z} = mv_{2x}r = -0.01 \text{ J s}$. Then the impulse components acting on C_2 are $J_{2x} = L_{22x} - L_{12x} = 0.0182 \text{ J s}$ and $J_{2z} = -0.01 \text{ J s}$. The complete impulse acting on C_2 is obtained from the sum of the squares of the component impulses, $J = (J_{2x}^2 + J_{2z}^2)^{1/2} = 0.021 \text{ J s}$. The opposite impulse acts on C_1 so that the new angular momentum components of C_1 are $L_{21z} = 0.0466$ and $L_{22x} = 0.0182 \text{ J s}$. The angle between the z-axis and the new angular momentum vector of C_2 is $\tan^{-1}(L_{22x}/L_{21z}) = 21.3°$.

12.3. The spinning top is shown in Fig. 12.16, in which its angle to the vertical is θ, the distance of the centre of mass from the contact with the floor is d, the weight of the disc is mg, and its radius is r. The top spins about its axis with an angular velocity ω and precesses with an angular velocity Ω. The rotating component of the angular momentum, L, of the top is $L \sin \theta$, and the torque which acts on the top is $mgd \sin \theta$. In this case

$$dL/dt = \Omega L \sin \theta = mgd \sin \theta$$

or

$$\Omega = \frac{mgd}{L} = \frac{mgd}{I\omega},$$

where I is the moment of inertia of the top (in effect, of the disc).

Since I may be written as the product of the mass, m, and the square of the radius of gyration, K, we could also write $\Omega = gd/K^2\omega$, showing that the mass of the top is irrelevant. For the disc, $K^2 = \frac{1}{2}r^2$, so that, using the values given, we have $\Omega = (9.8 \times 1.2)/0.2\pi = 1.87\,\text{s}^{-1}$ or 17.9 rpm.

12.4.   The system described in this example is an analogue of the spinning top and is treated in practically the same way. The torque acting on the assembly is the product of the total weight of the rotating assembly, acting through the centre of mass, and the horizontal distance of the centre of mass from the bearing, $\Gamma = (M+m)gl \sin \theta$, where $M+m$ is the mass of the system, l is the distance along the rod of the centre of mass from the bearing, and θ is the angle made by the rod with the vertical. The torque required to produce the precession with angular velocity Ω is $\Gamma = \Omega L \sin \theta$, where L is the precessing angular momentum. If I is the moment of inertia of the spinning disc and ω is its angular velocity, then $L = I\omega$, leading to $\Omega = [(M+m)gl]/I\omega$. Using the figures given and taking $I = \frac{1}{2}Mr^2$ produces an angular velocity of $\omega = 28.6\,\text{s}^{-1} = 273$ rpm.

12.5.   Since we treat the flywheel as a laminar disc, we can say that its principal axes are the Cartesian axes the yz-plane of which lies in the plane of the disc. The moments of inertia about the principal axes are then $I_x = \frac{1}{2}ma^2$ and $I_y = I_z = \frac{1}{4}ma^2$, where m is the disc mass and a is its radius. The angular momentum components for the disc are $L_x = I_x \omega_x$, $L_y = I_y \omega_y$, and $L_z = I_z \omega_z$, where ω_x, ω_y, and ω_z are the components of the angular velocity, $\boldsymbol{\omega}$, along the principal axes of the disc. The torque produced by the rotation of the angular momentum vector L is

$$\Gamma = dL/dt = \boldsymbol{\omega} \times L.$$

The components of Γ are

$$\Gamma_z = \omega_x L_y - \omega_y L_x = \omega_y \omega_x (I_y - I_x),$$
$$\Gamma_y = \omega_z L_x - \omega_x L_x = \omega_x \omega_z (I_x - I_z),$$

and

$$\Gamma_x = \omega_y \omega_z (I_z - I_y).$$

Since $\omega_z = 0$, $\Gamma_y = \Gamma_x = 0$, and $\Gamma_z = \frac{1}{4}ma^2\omega^2 \cos \theta \sin \theta$, which, for $v = 50\pi$, $a = 0.3$, and $m = 10$, has the value $\Gamma_z = 48.44\,\text{N m}$. The force acting at each bearing is $F = 48.44/0.1 = 484\,\text{N}$.

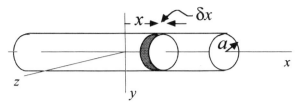

Fig. 12.17 The calculation of the principal moments of inertia of a cylinder with aspect ratio l/a.

One interesting point which arises from this example is that if the disc had a cylindrical shape—in other words, if we modelled the flywheel rather better—it's possible that the length of the cylinder could be adjusted to make the three principal moments of inertia identical. Then all three components of the torque would be zero, so that even when the axis of rotation of the flywheel is misaligned through its centre there would be no forces at the bearings due to the rotational motion. The critical value of the aspect ratio length/radius $= l/a$ for which this equality of the moments of inertia happens is easily calculated. The cylinder is drawn in Fig. 12.17. Obviously, $I_x = \frac{1}{2}ma^2$, just as if the cylinder were a lamina. I_y (and I_z) can be found by taking circular laminar disc elements of thickness δx, the moments of inertia of which about the z-axis, using the parallel axis theorem, are $(\frac{1}{4}a^2 + x^2)\mathrm{d}m = \pi a^2\rho(\frac{1}{4}a^2 + x^2)\delta x$, where ρ is the density of the material of the cylinder. On integrating with respect to x between $\frac{1}{2}l$ and $-\frac{1}{2}l$ we find that

$$I_y = \pi\rho a^2 l\left[\tfrac{1}{4}a^2 + \tfrac{1}{12}l^2\right] = m\left[\tfrac{1}{4}a^2 + \tfrac{1}{12}l^2\right].$$

Equating this to $I_x = \frac{1}{2}ma^2$ leads to the conclusion that $I_x = I_y = I_z$ when

$$l = a(3)^{1/2} = 1.732\,a.$$

We could perform a similar calculation for cross-sectional shapes (perpendicular to the disc face) of the circular flywheel other than the rectangular one dealt with here, but in each case there will be only one particular value of the aspect ratio for which the three principal moments of inertia are equal.

12.6. A cross-sectional diagram of this rotating disc/rotating table assembly is given in Fig. 12.18. If the radius of the disc is a, its mass m, and its angular velocity ω, then we may write its angular momentum as

$$L = I\omega = \tfrac{1}{2}ma^2\omega.$$

The vector L is parallel to the axis of rotation of the disc, so that rotation of the table about a vertical axis through its centre with angular velocity Ω requires a torque

$$\boldsymbol{\Gamma} = \boldsymbol{\Omega} \times \boldsymbol{L}$$

of magnitude $\Gamma = \Omega L$. The vector $\boldsymbol{\Gamma}$ (into the plane of Fig. 12.18) is itself

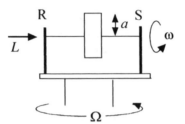

Fig. 12.18 A rotating disc mounted on a rotating table.

the vector product of a vertical force acting on the supports, R and S, in Fig. 12.18, and the half-length of the axle, $\boldsymbol{\Gamma} = \boldsymbol{d} \times \boldsymbol{F}$, so that $\boldsymbol{F}$ acts upwards on the left support, S, in Fig. 12.18, and downwards on the right-hand one, R. Because the vectors ($\boldsymbol{L}$, $\boldsymbol{\Omega}$, $\boldsymbol{\Gamma}$, $\boldsymbol{d}$, and $\boldsymbol{F}$) lie in the x-, the y-, or the z-directions of a suitably chosen set of axes—that is, they are at right angles to, or parallel with, the horizontal in Fig. 12.18—we may drop the vector notation to give

$$F = \frac{\Omega L}{d} = \frac{ma^2\omega\Omega}{2d}.$$

In the absence of rotation about the vertical axis the forces acting on the pillar supports R and S would be $\tfrac{1}{2}mg$ downwards. When the system is rotated the forces acting on the left- and right-hand supports will be

$$F_S = \tfrac{1}{2}m\left(g - \frac{a^2\Omega\omega}{d}\right), \qquad F_R = \tfrac{1}{2}m\left(g + \frac{a^2\Omega\omega}{d}\right).$$

Substituting in the values given leads to

$$F_S = \tfrac{1}{2}[9.8 - (0.01 \times 7.8 \times 4\pi)/0.1] = 0.294\,\text{N}, \qquad \text{and} \qquad F_R = 9.506\,\text{N}.$$

Inevitably, F_R and F_S sum to 9.8 N, so that the weight carried by the two supports is still the weight of the flywheel, but the distribution of the weight between the two pillars has changed radically. It's obvious, from the form of F_S, that for $\Omega = gd/\omega a^2$ the force acting on the left-hand post is zero and becomes negative as Ω increases beyond this critical value. When F_S is zero the left-hand support serves no purpose and could be removed; the flywheel would then assume a stable state despite its being supported at one end only. If Ω were increased beyond its critical value and the bearing was fixed in position, the left-hand pillar would be in tension and would have to be held down on the table.

The work done by a torque Γ acting through an angle θ is $\Gamma\theta$. We can assume fairly that the pillar supports are rigid, so that the torque resulting from rotation of the table will not produce any change in the position of the axle relative to the table; that is, $\theta = 0$. Provided that the bearing which carries the table is perfect, no work is required to rotate the table.

12.7. Take the radius of the earth to be R. The forces acting on a mass m at a

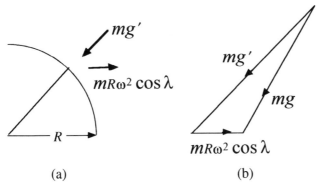

(a)   (b)

Fig. 12.19 (a) The forces acting on a body of mass m at a latitude λ on the earth's surface. (b) The force triangle to find the effect of the centrifugal force on the true gravitational acceleration, g'.

latitude λ, the gravitational force acting towards the earth's centre and the centrifugal force directed perpendicularly away from the axis of rotation, are shown in Fig. 12.19(a). The resultant of these two forces, shown in Fig. 12.19(b), is the apparent gravitational force, mg. Because the angular velocity of the earth's rotation is so small ($\omega = 7.3 \times 10^{-5} \text{ s}^{-1}$) the centrifugal force is small compared with the gravitational force and the angle $\lambda - \lambda'$ is very small. The evaluation of $\lambda - \lambda'$ is made by using the force triangle shown in Fig. 12.19(b). A version of this triangle is shown in Fig. 12.20. The line CD is drawn perpendicular to AB in the triangle and has a length of $mR\omega^2 \cos \lambda \sin \lambda$. Using the sine rule in the triangle BCD, we have

$$\frac{mg}{\sin(\frac{1}{2})\pi} = \frac{mR\omega^2 \sin \lambda \cos \lambda}{\sin(\lambda - \lambda')}$$

and, since $\lambda - \lambda'$ is small, we can substitute $\lambda - \lambda'$ for $\sin(\lambda - \lambda')$, to give

$$\lambda - \lambda' = \frac{R\omega^2 \sin \lambda \cos \lambda}{g}$$

$$= \frac{R\omega^2 \sin 2\lambda}{2g} = 1.73 \times 10^{-4} \sin 2\lambda.$$

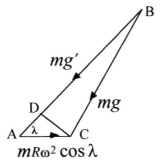

Fig. 12.20 The triangle of forces used to evaluate $(\lambda - \lambda')$ and g.

At 52°N, $\lambda - \lambda'$ has the value of 1.68×10^{-4}, and at the equator it is zero. Because $\lambda - \lambda'$ is so small we can make the approximation that the triangle BCD is to all intents an isosceles triangle. Then we may say that $AB = AD + DB = AD + BC$, and since $AD = mR\omega^2 \cos^2 \lambda$, the observed acceleration due to gravity, g, is related to the true gravitational acceleration, g', by

$$g = g' - R\omega^2 \cos^2 \lambda$$

or, in fractional terms,

$$\delta g/g = -(R\omega^2 \cos^2 \lambda)/g.$$

At the equator, with $\lambda = 0$, $\delta g/g$ is 0.034%, and at latitude 52°N it is 0.012%.

12.8. If the earth rotates with angular velocity $\boldsymbol{\omega}$ and the bolt's downward velocity is v, the Coriolis force acting on the falling bolt is $-2m(\boldsymbol{\omega} \times \boldsymbol{v})$, directed towards the east, where m is the mass of the bolt. At a latitude of $\lambda°$ this forces produces a 'transverse' acceleration, a_t, of magnitude $2\omega v \sin \lambda$. The bolt falls freely and so $v = gt$, where g is the acceleration due to gravity and t is the time of fall.

Then we have $a_t = 2\omega gt \sin \lambda$, $v_t = \int a_t \, dt$, and $x = \int_0^T v_t \, dt$, where T is the time for the bolt to fall to the ground. This leads to

$$x = \tfrac{1}{3}\omega g T^3 = \tfrac{1}{3}\omega g (2h/g)^{3/2},$$

where h is the height of fall. Here we have $h = 300$, so that $x = 8.6 \, \text{cm}$ to the east.

Would the centripetal force have any effect on the deviation of the path of the bolt from a radius vector of the earth? The first of the effects of the centripetal force, the reduction in magnitude of g, of the order of 0.03%, is negligible since x will be altered only by half that percentage. The second one, the angular deviation, $\lambda' - \lambda$, between the directions of g' and g, won't be significant because the tower will have been constructed using a vertical established by a plumb bob. The tower would be aligned with the vector g. If the tower had been constructed so that it were aligned with g' (along a radius vector of the earth) there would be a lateral deviation due to this cause. We could make an estimate of this lateral deviation of the point of impact due to the misalignment of g and g' by multiplying $\lambda' - \lambda$ by the height of fall, because $\lambda' - \lambda$ is a small angle. In this case $\lambda' - \lambda = 1.71 \times 10^{-4}$, and the lateral deviation would be $z = 5.1 \, \text{cm}$. Because z is in the direction due south, in this hypothetical case the point of impact would be 10 cm from the base of the tower in a direction 30°ESE.

Problems

1. A rugby-playing student has been selected for a game to be played on a pitch 2 km due north of his rooms. He walks to the playing field and, after the game, goes with his team-mates to a bar situated 3 km due east of the field. After a considerable lapse of time he sets off to return to his rooms, but his sense of direction is slightly impaired. When he wakes next day he finds himself lying on the grass in a public park some 10 km due south of his rooms. Construct a vector diagram of his peregrinations. What is the magnitude of the vector describing his misdirected late-night walk, and what is its direction?

2. A golfer having a bad day on the greens takes three putts to sink the ball into the hole. The first putt is of 12 m due west, the second of 6 m due NE, and the third of 3 m due SE. How far from the hole was the ball before the first putt? In what direction should the putt have been made for the golfer to hole out with a single shot?

3. A pedantic student of physics determines to map out the campus of his university in terms of vectors. He chooses unit vectors **i** in a direction due east, **j** in a direction due north, and **k** in a direction vertically upwards. Naturally, he takes the position of the Department of Physics as origin. If the library building is 250 m from the department in a direction W21°N, and on the same level, what are the components of the vector joining the library to the department in his chosen coordinate system? If the books in the physics classification are in the library basement 20 m below ground level, what are the components, and what is the magnitude, of the vector joining the department to the shelves carrying the physics books in the library?

4. From its launch, a hot air balloon travels 600 m due SW until its altitude is 400 m, and then travels 800 m due SE while climbing through a further 200 m. If this has taken 15 min from the time of launch, what is the average speed of the balloon?

5. An oil tanker is sailing due east across a current which flows due south. In a quarter of an hour it has travelled 6 km in a direction 15°ESE. What are the speeds of the current and of the ship?

6. The lookout on a whaling ship, A, sailing due west at 12 kph, spies a school of whales, B. If the whales are swimming at 16 kph due south, what is the relative velocity of the whales with respect to the ship?

7. A passenger aircraft has an average air speed of 720 kph and a maximum range of 5760 km (in still air) when it is carrying 180 passengers. This range is

increased by 18 km for each empty seat in the aircraft. The aircraft is to fly to a destination at its maximum range but will fly into a headwind of 90 kph. How many passengers can it carry and still reach its destination safely?

8. A cannon is mounted on a train travelling due north with a constant velocity V. If the barrel of the cannon is set at an angle ϕ due E of N, and the muzzle velocity of a shell it fires is v, what is the direction of the path of a shell fired horizontally from the cannon?

9. A racing cyclist in a time trial travels along a straight road running north–south at a steady speed of 48 kph and notices that the wind direction is N40°E when he is going north. Returning along the same road at the same speed he observes that the wind appears to be coming from S30°E. What is the wind speed, and what is its direction?

10. A batsman is at the wicket, W, and a fielder is in the outfield at F. The batsman strikes the ball in a direction making an angle of 30° with the line WF, with a speed 1.5 times that with which the fielder can run. If the fielder starts off straightaway at his maximum speed, determine the direction in which he must run in order to field the ball as soon as possible. Show that if in doing this he has run 20 m, then he was standing about 39 m from the wicket.

11. Huck and Tom have a race crossing a river to a post at a point on the far bank directly opposite the point on the near bank at which the race started. Both can swim at 5 kph, the river is 150 m wide, and the current in the river is 8 kph. They adopt different strategies—Huck decides to swim in the direction which will get him to the opposite bank in the shortest possible time and Tom swims in the direction which will take him as little downstream as possible. How long will they take to cross the river? If Huck can swim at $5\,\mathrm{m\,s^{-1}}$ and Tom at $8\,\mathrm{m\,s^{-1}}$, who will win the race?

12. A passenger ferry is equipped with a radar set which has an effective range of 25 km. It leaves harbour and steams due north at 16 kph. At the time it sails, another ship is 75 km due NE of the port, following a course due west. This second ship will be detected by the ferry's radar only if its speed is between a lower and an upper limit. What are these limits? If the ferry sails at midnight, what are the earliest and latest times its radar will detect the second ship, given that its speed is between the two limiting values for detection?

13. Two cars, P and Q, with speeds of 30 and 40 kph respectively, are travelling along straight roads directed at 90° to one another towards a crossroads at R. If PR is 750 m when QR is 1200 m, what is the shortest distance between the two cars as they proceed on their journeys?

14. Two aircraft, A and B, are approaching an airfield, one with a velocity of 200 kph due S of E and the other with a velocity of 200 kph due north. If A is observed to be 30 km from the airfield when B is 20 km from it, what will be the distance of closest approach? When the aircraft are closest together, how far from the airfield will aircraft A be?

15. A cruise ship is travelling at 30 kph on a straight course which will take it past a port with a distance of closest approach of 6 km. When the ship is 10 km from the port, but has not yet reached its point nearest to the port, a tugboat leaves the port to intercept it. What is the minimum uniform speed the tug must have so that the interception can take place? If the tug travels at 24 kph, for how long a time will the cruise ship be on that part of its course where the tug can intercept it? If the closest approach of the ship to the port occurs when the ship is due north of the port, what are the limits on the course of the tug travelling at 24 kph for which interception will take place?

16. A car travels at u kph along a road which is crossed by a lane at right angles to its direction. Show that it is impossible for a man in the car to throw an object straight down the lane unless the speed, v, with which he can throw is greater than u. Find the greatest acute angle between the lane and the road for which this feat is possible if $u = 40$ kph and $v = 25$ kph.

17. An aircraft has a take-off speed of $80 \, \mathrm{m \, s^{-1}}$ and accelerates in its take-off run at $4 \, \mathrm{m \, s^{-2}}$. What length of runway does the aircraft require to get off the ground? If the pilot aborts the take-off just as flying speed is attained, and the runway is 1500 m long, what is the minimum retardation that would be required to stop the aircraft on the runway?

18. Our competition sports car has a maximum speed of 252 kph along the straight and of 180 kph around a bend of radius 0.5 km. The race track on which it is running consists of two 4 km straights joined by two semicircular curves of radius 0.5 km. If the maximum acceleration and retardation of the car are both $4 \, \mathrm{m \, s^{-2}}$, and the current lap record is at an average speed of 220 kph, will we be able to set a new lap record; and, if so, what would it be?

19. An electric tram car travelling between two stops 400 m apart is uniformly accelerated for the first 10 s, during which time it travels 30 m. It then runs with uniform velocity until it is 15 m from the second stop, when it is retarded uniformly to come to a halt at the stop. How long has the journey taken?

20. The gun carriages of cannons on eighteenth-century men-of-war were restrained in their recoil by a rope and pulley harness. Such a cannon, of mass 1 tonne, loaded with a cannonball of mass 5 kg, was fired from the level deck of the *U.S.S. Constitution*. If the velocity of the cannonball as it left the muzzle of the gun was $400 \, \mathrm{m \, s^{-1}}$ in a horizontal direction, what was the initial velocity of recoil of the cannon? If the harness brought the gun to rest after it had moved back from its initial position by 40 cm, what was the average force exerted by the harness on the cannon?

21. A pile driver of mass 5 tonnes is used to drive a foundation pile of mass 1 tonne into the ground. The pile driver falls through a height of 3 m on to the pile. If the pile is driven 5 cm into the ground, what is the average resistive force exerted by the ground?

22. During a heavy tropical rainfall, 10 cm of rain fell in 2 h. If the velocity of the rain when it struck the ground was equal to that it would have attained by

falling freely through $300 \mathrm{m}$ and the density of water is $10^3 \mathrm{kg\,m}^{-3}$, what was the average pressure exerted on the ground by the rain?

23. A jet of water rises vertically at a speed of $10 \mathrm{ms}^{-1}$ from a nozzle of cross-sectional area $0.6 \mathrm{cm}^2$. A ball of mass $0.5 \mathrm{kg}$ is balanced in air by the impact of the water on its underside. What is the height of the ball above the nozzle?

24. The effect of air resistance on the motion of an automobile can be specified by the effective area of the car against resistance. If an effective area of $1.5 \mathrm{m}^2$ can be ascribed to a vehicle which is travelling against a headwind of $16 \mathrm{kph}$ at a speed of $100 \mathrm{kph}$, how much power is used in overcoming wind resistance? The density of air is $1.25 \mathrm{kg\,m}^{-3}$.

25. What power would be required to make a horizontal jet of water issue from an orifice of area $64 \mathrm{cm}^2$ at a rate of $4.8 \mathrm{m}^3 \mathrm{min}^{-1}$? If the jet were directed against a vertical wall, what force would it exert on the wall?

26. A railway train has a mass of $200 \mathrm{tonnes}$. If the resistive force opposing its motion when it travels at $100 \mathrm{kph}$ is $10 \mathrm{kN}$, what will be the power required to maintain its speed at $100 \mathrm{kph}$?

27. A compact car of mass $1 \mathrm{tonne}$ accelerated uniformly from rest along a level road attaining a velocity of $18 \mathrm{kph}$ after $10 \mathrm{s}$. If the forces resisting its motion are constant at $50 \mathrm{kN}$, what power is expended by the car as it travels at $18 \mathrm{kph}$?

28. The engine of a motor cycle working at a steady rate of $5.6 \mathrm{kW}$ drives the bike at a uniform speed of $80 \mathrm{kph}$ against a constant resistive force. If the mass of the bike (and its rider) is $150 \mathrm{kg}$, at what speed will it climb a hill of 1 in 10 if its engine produces the same power and the resistive forces are unchanged?

29. In a nineteenth-century steam engine the diameter of the low pressure cylinder was $1.2 \mathrm{m}$, the average speed of the piston was $4 \mathrm{ms}^{-1}$, and the average pressure of the steam acting on the piston was $2.4 \times 10^5 \mathrm{Pa}$. Make an estimate of the power of the engine.

30. What power is required to pump $4.5 \mathrm{m}^3$ of water per minute from a depth of $15 \mathrm{m}$ and deliver it through a pipe of cross-sectional area $40 \mathrm{cm}^2$? The density of water is $1 \mathrm{tonne\,m}^{-3}$.

31. A piece of armour plate is made by bonding a $4 \mathrm{cm}$ thickness of plastic on to a $2 \mathrm{cm}$ thick steel plate. In a test, a bullet of mass $50 \mathrm{g}$ is fired with a velocity of $400 \mathrm{ms}^{-1}$ at the plastic side of the armour. It penetrates through the plastic and embeds itself $1 \mathrm{cm}$ into the steel. When the bullet is fired at the steel side of the armour it penetrates the steel and embeds itself $2 \mathrm{cm}$ into the plastic. Assuming that uniform forces are exerted on the bullet by the two materials, what is the ratio of the forces on the bullet due to the steel and to the plastic? What thickness of plastic would be required to stop the bullet in plastic?

32. An elevator leaves the ground floor of a tall building with an acceleration of $4 \mathrm{ms}^{-2}$ and slows down with a deceleration of $4 \mathrm{ms}^{-2}$ as it approaches the

top floor of the building. If the only passenger in the elevator cage has a mass of 70 kg, to what force is the floor of the cage subjected during its ascent? At the top floor six passengers of total mass 500 kg enter the elevator. If it is subject to the same accelerations (i.e. of the same magnitude but of opposite sign) during its descent as it was during its ascent, what forces act on its cage floor as it descends?

33. An aircraft is flying at 1500 kph. When it turns in a horizontal circle it does so without side slipping. If its wings will collapse (fold up) when they are subjected to a force greater than ten times its weight, what is the minimum radius of horizontal turning circle in which it can fly at this speed? What is the angle of bank in such a turn?

34. When it is struck by the bat a baseball of mass 250 g attains a velocity of $37 \, \text{ms}^{-1}$. The ball flies direct to a fielder some 20 m from the plate, who catches it as the same level at which it was hit. He observes that its velocity has been reduced to $34 \, \text{ms}^{-1}$. Make an estimate of the average drag force (the force of air resistance) assuming that the reduction in velocity is due entirely to air resistance.

35. A frisbee of mass 100 g is thrown from a height of 1 m with a velocity of $12.5 \, \text{ms}^{-1}$. When it has reached a height of 2.3 m its velocity has decreased to $9.6 \, \text{ms}^{-1}$. How much energy has it lost to air drag?

36. When it is burned, 1 kg of good quality coal produces 35 MJ of energy. The velocity of the earth in its orbit is about $30 \, \text{kms}^{-1}$. If the earth were made entirely of coal, would more energy be produced by burning it or by stopping it in its orbit?

37. World-class pole vaulters regularly clear 5.5 m. Assuming that the centre of mass of a vaulter of mass 80 kg is 1 m above ground level and that his kinetic energy in the run-up is converted entirely to potential energy in the vault, how fast would he be running when he entered his vault?

38. A spring is attached to a hook in the ceiling and is loaded with a weight attached to its free end. If the weight is lowered gently to its equilibrium position, by how much does its potential energy change? Is this the same as the change in potential energy of the spring; and if not, why not?

39. The interaction between two nucleons (protons or neutrons) in a nucleus can be written, for separations of more than $10^{-15} \, \text{m}$ ($= 1 \, \text{fm}$), in terms of the potential energy of interaction, as the Yukawa potential, $V(r) = -[\lambda \exp(-r/r_0)]/r$. r is the separation of the nucleons, r_0 is taken as $1.4 \times 10^{-15} \, \text{m}$, i.e. 1.4 fm, and $\lambda = 70 \, \text{MeV fm}$ ($1 \, \text{MeV} = 1.6 \times 10^{-19} \, \text{J}$). What is the internuclear force acting on two protons separated by 1.4 fm?

40. A truck of 5 tonnes mass pulls up sharply at a red light? The driver of the following truck suffers a lapse in concentration and runs, at $2 \, \text{ms}^{-1}$, into the back of the static truck. If the mass of this second truck is 10 tonnes and the relative velocity of the trucks after the impact is $0.6 \, \text{ms}^{-1}$, how much energy has been lost in the collision?

41. A snooker ball, of mass 213 g, with a velocity of 2.83 m s^{-1}, which is inclined at an angle of 45° to a smooth cushion at the edge of the snooker table, collides with the cushion. If the coefficient of restitution is 0.875, what is the loss of kinetic energy on impact?

42. A mortar is trained from a trench on to another trench at the same level, 1 km distant. What should the angle of elevation be if the bomb, when fired with a velocity of 125 m s^{-1}, is to hit its target? What is its time of flight?

43. A car of mass 0.9 tonne has an engine with a maximum power output of 22.2 kW. The resistance to its motion has the form $(kv^2 + 200)$ N, where v is its velocity and k is a constant. If its maximum speed on the level is 30 m s^{-1}, what is the value of k? How much power would the engine need to develop if the vehicle were to maintain a speed of 24 m s^{-1} up a slope of 1 in 16?

44. After a sharp frost, a child discovers a storage drum of 2.0 m diameter lying on its side. He places a pebble on its topmost point and allows it to slide down the side of the drum. At what height from the ground will the pebble leave the drum? At what position will it hit the ground?

45. In a film about the declining years of the composer Delius, Percy Grainger was shown throwing a ball over the Delius residence, rushing through the house, and catching it before it landed in the back garden. If Grainger had to run a distance of 24 m from the point of projection with an average speed of 8 m s^{-1} in order to make a clean catch, what was the angle of projection of the ball, and what was the velocity with which he threw it?

46. A seaside cable railway enables holidaymakers to ascend from sea level to an hotel 20 m higher without effort on their part. If the car on this railway has a mass of 1 tonne and is accelerated at 0.5 m s^{-2} up a slope of 1 in 2 against frictional forces which are characterized by a coefficient of friction $\mu = 0.1$, what is the tension in the cable?

47. A capacitor of capacity 1 nF is charged by connecting it to an electric cell, the e.m.f. of which is 1 V. The capacitor is then isolated from the cell and connected across an inductor the inductance of which is 10 μH and the resistance of which is negligibly small. What is the form of the current in the circuit?

48. The newly discovered planet Bacchus has a mass M and a radius b. If it had a single satellite of mass m held in its orbit of radius s by gravitational attraction, show that the period of one orbit of this moon (satellite) could be expressed in terms of the acceleration, γ, due to gravity at the surface of Bacchus as $\mathcal{T} = 2\pi(s^3/\gamma b^2)^{1/2}$. In fact, Bacchus has three moons, each of mass m, named X, Y, and Z, all in this same orbit of radius s, and their positions are such that the triangle XYZ is always equilateral. How does the period of revolution of X (or Y, or Z) compare with $\mathcal{T}$?

49. A spring balance carries a load of 6 kg. Its user hangs a 1.5 kg bag of flour as an additional load and observes that the spring extends a further 5 cm. After a minute or so the flour bag tears and the flour is deposited on the floor. This

sudden removal of the flour sets the 6 kg load into oscillation. What is the period of this oscillation? When the load is 2.5 cm above its lowest point what is its velocity, and what is the tension in the spring?

50. The acceleration due to gravity is taken to vary inversely with the square of the distance from the earth's centre. A pendulum clock which keeps correct time at sea level is taken to an elevation of 1.6 km when its owner moves house from the seaside. Taking the radius of the earth as (6.4×10^3) km, what percentage change must be made in the length of the pendulum for the clock to keep correct time in its new environment?

51. During a severe earthquake a table top oscillates horizontally with simple harmonic motion of amplitude 6 cm. It completes 60 oscillations in one minute. If a parcel is resting on the table, what is the minimum coefficient of friction required to keep it in its place on the table?

52. A hatch in a vertical wall is closed with a trap door of uniform thickness which is hinged along its top edge. The door is 80 cm wide, 1.2 m high, and of mass 20 kg. When it is open, it is held in a horizontal position by a catch. When the catch is released the door closes by swinging shut under the influence of gravity. What is the velocity of the bottom of the door when it strikes the wall?

53. An enthusiastic lecturer makes a large demonstration gyroscope using a disc with a fixed axle (through its centre, perpendicular to its plane) mounted on a set of gimbals. He sets the wheel in motion by winding a cord of 3 m length around the axle and pulls it with a constant force of 100 N. When the cord leaves the axle the rod is rotating at 600 rpm. What is the moment of inertia of the wheel/axle combination?

54. A callow student mistakenly believes that he has made a simple pendulum of length 1 m when he suspends a 1 kg mass at the end of a metre rule of mass 150 g and swings it to perform small oscillations about its unloaded end. What will be the difference between the period of oscillation he measures and the period of oscillation of a 'proper' 1 m long simple pendulum? Assuming that he uses his value of the period of oscillation to calculate a value for the acceleration due to gravity, what error will be introduced by his incorrect assumption?

55. A cyclist has turned his cycle upside down to repair a puncture. Out of curiosity he sets the rear wheel spinning at 200 rpm and observes that it comes to a stop after 90 s. Assuming that the decelerating torque is constant, how many revolutions does the wheel make before stopping? If the wheel has a diameter of 60 cm and its weight of 1 kg can be taken as concentrated all at its rim, what is the frictional torque that stops the wheel?

56. A power press for stamping metal sheet uses the energy stored in a flywheel driven by a motor with a power of 4 kW. Each 10 s it forms a sheet and the process absorbs 80% of the energy supplied in that time. If the rate of revolution of the flywheel can vary only from 130 to 100 rpm and its mass is 2 tonnes, what is the smallest radius allowed for the flywheel?

57. A railway coach door has a vertical hinge and a width of 0.8 m. Usually an employee of the railroad company closes the door before the train departs, but for once this task is not completed. If the lock on the door will engage only when the angular velocity of the door exceeds $2\pi/3\,\mathrm{s}^{-1}$, what uniform acceleration of the train will cause the door to close if its initial position was at right angles to the side of the train and the hinges allowed it to close?

58. The central spindle of a yo-yo has a diameter of 2.5 cm and at each of its ends it has a cylindrical disc of diameter 5 cm. The spindle is 5 mm long, the 'capping' discs are each 10 mm in width, and the density of the material from which the toy is made is $2\,\mathrm{tonnes\,m}^{-3}$. If the loose end of the yo-yo string is held in position and the yo-yo falls under the influence of gravity, what is the tension in the string?

59. It is conjectured that the builders of the pyramids moved their stone blocks by placing rollers under one of the stones and pushing it along. Let's say that one such stone had a mass of 10^3 tonne and rested on three rollers each of mass 100 kg and diameter 30 cm. If the stone was pushed on its rollers along a horizontal track at a uniform rate equivalent to $200\,\mathrm{m\,hr}^{-1}$ relative to the ground, what was the kinetic energy of the stone/roller system?

60. The lecturer of Problem 53 decided to make a more manageable version of his demonstration gyroscope. On a table which was free to rotate about a vertical axis, he attached two identical spring balances, equidistant from the centre of the table along the same line passing through the centre of the table. He adapted the balance pans to act as bearings for a flywheel and axle with radii of 5 cm and 2 mm, and with masses of 120 g and 20 g, respectively. If he could spin the flywheel at 900 rpm, and the axle was 5 cm long, at what rate did he have to rotate the table about its vertical axis so that one of the spring balances read zero?

Answers to problems

1. The track of the student is the sum of the paths $2\mathbf{j}$, $3\mathbf{i}$, and the unknown path $r = x\mathbf{i} + y\mathbf{j}$. Since this takes him to the point $-10\mathbf{j}$,

    $$- 10\mathbf{j} = (x + 3)\mathbf{i} + (y + 2)\mathbf{j},$$

    giving

    $$r = - 3\mathbf{i} - 12\mathbf{j}, \qquad r = 12.37 \, \text{km along } S14.1°W.$$

    This can be compared with the distance 3.61 km along W33.7°S that he should have travelled.

2. Taking the initial position of the ball as origin, O, we can draw the track of the ball as shown in Fig. P.1. Taking the E−W axis as the x-axis with unit vector, $\mathbf{i}$, we can describe the path of the ball after the first stroke as $r_1 = -12\mathbf{i}$, where $|\mathbf{i}| = 1 \, \text{m}$. The second and third strokes produce the paths $PQ = r_2$ and $QH = r_3$ with $r_2 = (6\cos 45°)\mathbf{i} + (6\sin 45°)\mathbf{j}$ and $r_3 = (3\cos 45°)\mathbf{i} - (3\sin 45°)\mathbf{j}$, with $\mathbf{j}$ the unit (1 m) vector in the direction of north. On enumerating the trigonometric functions we obtain the three equations

    $$r_1 = - 12\mathbf{i}, \qquad r_2 = 4.243\mathbf{i} + 4.243\mathbf{j}, \qquad r_3 = 2.121\mathbf{i} - 2.12\mathbf{j},$$

    so that

    $$r_1 + r_2 + r_3 = - 5.646\mathbf{i} + 2.12\mathbf{j}.$$

    The distance OH is given by $|r|^2 = (5.646^2 + 2.121^2)^{1/2}$, so that

    $$OH = |r| = 6.02 \, \text{m}.$$

    The angle required to define r fully is $\theta = \tan^{-1}[2.121/(-5.636)] = 159.4°$ or 21.6° N of W.

    Assuming that the hole is about 10 cm in diameter and that the green is

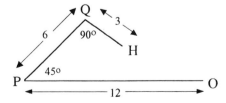

Fig. P.1 The track of the golf ball from its initial position, O, on the green, to the hole, H, by way of the three shots OP, PQ, and QH.

billiard table smooth, the ball would be holed provided that it was directed within ±0.5° of this bearing from the initial 6 m position. On the other hand, the fact that the initial putt missed the hole by more than 2 m implies that the golfer misread the humps and hollows of the green, which wasn't flat at all.

3. The department–library vector has the components

$$r_x = -250\cos 21° (= 250\cos 159°) = -233.4$$

and

$$r_y = 250\sin 21° (= 250\sin 159°) = 89.6.$$

For the bookshelf,

$$r_x = -233.4, \qquad r_y = 89.6, \qquad r_z = -20,$$

so that

$$|r| = \sqrt{(250^2 + 20^2)} = 250.8.$$

4. The magnitude of the average velocity of the balloon is given by its distance from the launch point divided by the time taken to travel that distance. Take the unit vectors in the SW and SE directions to be **i** and **j** respectively, and the unit vector in the vertical direction to be **k**. Using the launch point as origin, O, in Fig. P.2, the two legs of the balloon's path are

$$r_1 = 600\mathbf{i} + 400\mathbf{k}, \qquad r_2 = 800\mathbf{j} + 200\mathbf{k},$$

so that the position of the balloon after 15 min is

$$r = r_1 + r_2 = 600\mathbf{i} + 800\mathbf{j} + 600\mathbf{k}.$$

The distance of the balloon from O is given by

$$|r|^2 = (600^2 + 800^2 + 600^2) = 1.34 \times 10^6,$$

so that $r = 1158$ m. The speed of the balloon is $1.29\,\mathrm{m\,s^{-1}}$, or 4.63 kph. From the triangles OAB and OBC in Fig. P.2 it's straightforward to show that the balloon lies 5.14° S of E from its launch point and that the angle of its elevation from O is 30.96°. The average velocity of the balloon is thus 4.63 kph along a line which is inclined at 31° to the horizontal in a direction 5.1° S of E.

The identification of speed with the magnitude of the average velocity can be misleading in other than one-dimensional motion. As an example, say that

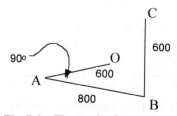

Fig. P.2   The track of the balloon.

we travelled between two towns lying 50 km apart on a straight inter-state highway. Instead of taking the direct route we went by the scenic route involving two straight legs of length 30 and 40 km, along which we drove at 60 and 40 kph respectively. What is our average velocity, and what is our average speed? The times taken on the two legs are 0.5 h and 1 h, so that our average velocity is 50/1.5 = 33.3 kph. We would estimate our average speed by dividing the distance we travelled by the time taken; that is 70/1.5 = 46.7 kph. The average velocity is less than both the speeds that we maintained in our journey, and less than our average speed.

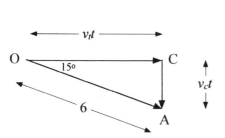

Fig. P.3 The resultant velocity vector OA of the ship and its components.

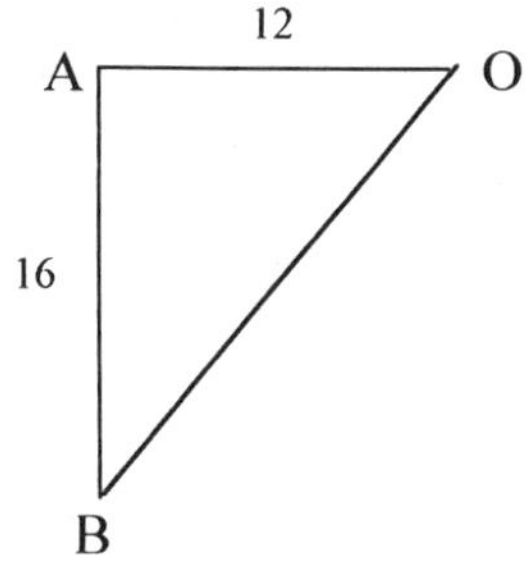

Fig. P.4 The velocity vector diagram for the whaling ship and the whales.

5. The track of the tanker is represented by a vector OA in Fig. P.3. The components of this vector in the directions south and east are OC and CA respectively. The angle COA is 15°, so that $OC = 6 \cos 15° = 5.8$ km and $CA = 6 \sin 15° = 1.55$ km. Since it takes the tanker 0.25 h to travel from O to A, the velocities of the tanker and the current are 23.2 and 6.2 kph respectively.

6. In Fig. P.4, OA and OB represent the velocities of the ship and the whales respectively. The side AB which closes the velocity triangle is the velocity of the whales with respect to the ship. The length of AB is given by $AB^2 = 16^2 + 12^2 = 400$, so that the velocity of the whales relative to the ship is 20 kph at an angle of $\tan^{-1}(4/3) = 53.1°$ S of E.

7. The duration of the flight when there is no wind would be 5760/720 = 8 h. Against the headwind the duration of the flight would be 5760/630 = 64/7 h. As far as fuel consumption is concerned, the extra distance to be travelled is $720[(64/7) - 8] = 823$ km. This is equivalent to 45.7 empty seats, so that the maximum safe load will be 134 passengers.

8. When the shell leaves the muzzle of the cannon it has a velocity v relative to the muzzle, which can be considered as made up from the components $v \cos \phi$ due north and $v \sin \phi$ due east. For a static observer at the side of the track the northerly component of the shell velocity is the sum of V and $v \cos \phi$, while the easterly component is $v \sin \phi$ (the same for the static observer and

for the artilleryman on the train). The angle θ, due E of N, at which the shell travels relative to the track, is given by $\tan \theta = [v \sin \phi / (V + v \cos \phi)]$.

As far as the artilleryman on the train is concerned, the shell leaves the cannon at an angle ϕ due E of N and the surrounding countryside is running past him with a velocity V due south. His frame of reference is travelling at this rate due north with respect to the track. The trackside observer sees the cannon travelling due north with velocity V; his frame of reference is static. The angle θ defining the direction in which the shell will travel is measured in this static frame. If the artilleryman was aiming at a target at a bearing ϕ from the train, he would have to lay off his aim in order to hit it. Because of the large discrepancy between V and v (probably V is only about 1% of v) we might be tempted to say that the correction is negligible, much less than 1°. On the other hand, if the range of the shell is R, neglecting the correction would lead to a miss by about $R\,(\theta - \phi)$, which will be between 15 and $20\,\mathrm{m\,km}^{-1}$ for $\theta - \phi = 1°$. Additionally, since the range can be taken as proportional to the square of the muzzle velocity, a range correction will need to be made, since the shell has a velocity W in the static frame, given by $W^2 = v^2 \sin^2 \phi + (V + v \cos \phi)^2 = v^2 + V^2 + 2Vv \cos \phi$. For $v/V = 1\%$, the range correction will be maximum at about 2% for $\phi = 0$, reducing to about $10^{-2}\%$ for $\theta = 90°$, and then increasing to about 2% for $\phi = 180°$ (2% is $20\,\mathrm{m\,km}^{-1}$).

9. In velocity space there are two vector triangles which have a common side because the wind stays constant in speed and direction. This situation is illustrated in Fig. P.5. The wind speeds observed by the cyclist are obtained by applying the sine rule to the 'double' triangle,

$$\frac{96}{\sin 110°} = \frac{a}{\sin 30°} = \frac{b}{\sin 40°},$$

$$a = 51.08, \qquad b = 65.67.$$

The cosine rule is used to give the wind speed, as

$$w^2 = 48^2 + 65.67^2 - (96 \times 65.67) \cos 30°$$

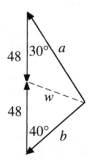

Fig. P.5 The velocity vector diagrams for the two legs of the time trial with constant wind *w*.

or $w = 34.01$, and the direction of w is found from the sine rule applied to one of the component triangles; for example,

$$\frac{a}{\sin \theta} = \frac{w}{\sin 30°} ,$$

leading to a wind direction of E15.12°S.

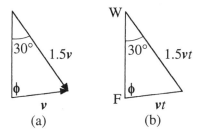

(a)   (b)

Fig. P.6   The (a) velocity and (b) space vector diagrams for the ball−fielder problem.

10.   The velocity and distance vector diagrams for the ball and the fielder are given in Fig. P.6. From the space diagram, using the sine rule,

$$1.5v \sin 30° = v \sin \phi$$

where v is the fielder's running speed, giving $\sin \phi = 0.75$ and $\phi = 48.59°$.
   The distance the fielder has run is vt, i.e. $vt = 20\,\text{m}$. The ball has then travelled 30 m and the length FW is obtained from the cosine rule. The third angle of the triangle is 101.41°, so that

$$(FW)^2 = 20^2 + 30^2 + (40 \times 30 \times \cos 78.59°)$$

or

$$FW = 39.2\,\text{m}.$$

11.   The velocity vector diagram for a swimmer in the river is shown in Fig. P.7, in which OA and OB represent the velocities of the river and the swimmer respectively. The sum of the two velocities will have a magnitude given by

$$v^2 = 8^2 + 5^2 + 80 \cos \theta$$

and the component of this velocity at right angles to the flow of the stream will be

$$v_{\perp} = 5 \cos(\theta - \tfrac{1}{2}\pi) = 5 \sin \theta$$

which is a maximum for $\theta = \tfrac{1}{2}\pi$ (Huck's strategy), with a resultant velocity of magnitude 9.43 kph. The time taken to cross the river with this velocity is $(0.15/5) \times 60 = 1.8\,\text{min}$. In this time Huck will have travelled $(1.8 \times 60) \times 8\,\text{km}$ downstream, i.e. 0.24 km or 240 m. Since he can run at $5\,\text{ms}^{-1}$ he will take a further 48 s to complete the race. His total time is 2 min 36 s.
   To find the downstream displacement a position vector diagram may be drawn as in Fig. P.8 in which OA represents Tom's course and OB represents his track. Tom's downstream displacement will be $y = (8t - 5t \cos \theta)$, where t is the time that he takes to cross: t may be calculated from the ratio of the

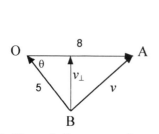

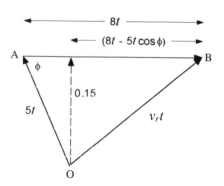

Fig. P.7 The velocity vector diagram for the swimmer in the river.

Fig. P.8 The displacement vector diagram for the swimmer in the river.

width of the river as $t = (0.15/5)\sin \phi$, so that we can substitute for t in the equation for y, to give y as a function of ϕ only, i.e.

$$y = \frac{0.15}{5 \sin \phi}[8 - 5 \cos \phi] = 0.15 \left[\frac{8}{5} \frac{1}{\sin \phi} - \cot \phi \right].$$

As y will be a minimum for $dy/d\phi = 0$, on differentiating,

$$\frac{dy}{d\phi} = 0.15 \left[-\frac{8}{5} \frac{\cos \phi}{\sin^2 \phi} + \cosec^2 \phi \right]$$

$$= \frac{0.15}{\sin^2 \phi}[1 - \tfrac{8}{5} \cos \phi] = 0,$$

leading to y a minimum for $\cos \phi = \tfrac{5}{8}$. This then gives

$$t = (0.15/5 \sin \phi) = 0.0304 \, \text{h, or } 2.3 \, \text{min,}$$

and

$$y = 0.0384[8 - (25/8)] = 0.187 \, \text{km, or } 187 \, \text{m.}$$

Tom attempts to swim in a direction which is inclined at 51.3° to the bank in the upstream direction. His time to the post will be $[2.3 + (187/\{8 \times 60\})] = 2.69$ min. Huck wins the race by about 5.5 s.

12. To avoid radar contact, the distance of closest approach must be more than 25 km. This leads to the kite-shaped track diagram shown in Fig. P.9, in which the angle on either side of NE is given by

$$\tan \theta = (25/75), \qquad \theta = 18.45°.$$

The velocity vector diagram is a right angled triangle because the courses of the two ships are at 90° to each other. One side of this velocity triangle is 16, and the angle between it and the hypotenuse is $(18.43 + 45)°$ or $(45 - 18.43)°$, i.e. 63.43° or 26.57°. These lead to relative velocities of 35.8 and 17.9 knots respectively. The corresponding velocities for the second vessel are 32 and

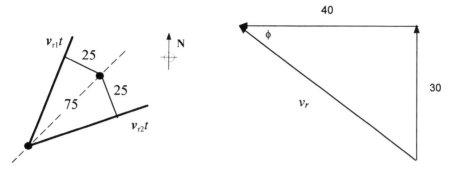

Fig. P.9 The representation of the two possible cases of a closest approach of 25 km.

Fig. P.10 The velocity vector triangle for the two automobiles.

8 knots. If its speed lies between these limits it will appear on the radar screen of the ferry. If its speed is slower than 8 knots, or greater than 32 knots, it will not be detected.

The time limits are 0159 and 0357.

13. The velocity vector triangle for the two cars is shown in Fig. P.10, from which it is seen that the relative velocity, $v_r = 50$ kph inclined at an angle of $\sin^{-1} 0.6$ to the direction of the velocity of Q (the horizontal in the diagram). The position vector diagram is shown in Fig. P.11, in which AC and EC represent the initial positions of P and Q respectively. C represents the position of the crossroads. The line AD is drawn parallel to the direction of v_r and represents the track of P with respect to Q. The line DE is the distance of closest approach, with the angle BDE a right angle. The length of AB is $0.75/\sin \phi = 1.25$ km, so that $BC = 1$ km. The distance of closest approach $DE = BE \sin \phi = 0.2/0.6 = 0.12$ km, or 120 m.

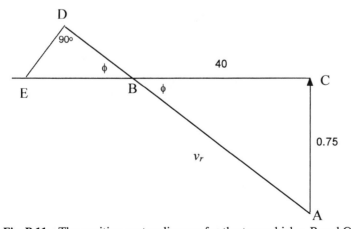

Fig. P.11 The position vector diagram for the two vehicles, P and Q.

14. The velocity vector diagram for the two aircraft, shown in Fig. P.12, is an

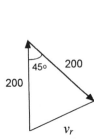

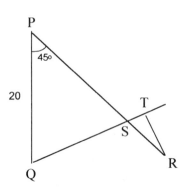

Fig. P.12 The velocity vector diagram for the two aircraft.

Fig. P.13 The position vector diagram for the two aircraft.

isosceles triangle with an apex angle of 45° so that the magnitude of the relative velocity $v_r = 2 \times 200 \sin 22.5° = 153.1$ kph. The relative velocity vector is inclined at an angle of 67.5° to the vertical in the diagram so that its direction is 67.5° E of N. The position vector diagram for the two aircraft (Fig. P.13) contains an isosceles triangle which is similar to the velocity vector triangle, so that PS = 20 km and SR = 10 km. The distance of closest approach is RS sin 67.5° = 9.24 km.

In passing, it's easy to calculate the positions of the aircraft relative to the airfield when they are at the distance of closest approach to one another. In the position vector diagram, the time at which closest approach occurs is given by $QT/v_r = (2 \times 20 \sin 22.5° + 10 \sin 22.5°)/(400 \sin 22.5°) = 0.125$ h. The time taken by aircraft B to arrive at the airfield is 0.1 h, so that its distance from the airfield when closest approach between the aircraft occurs is $0.025 \times 200 = 5$ km.

15. The initial position of the ship relative to the port, the ship's course, and the distance of closest approach are shown in Fig. P.14 as CP, AC, and AP respectively. For AP to be the distance of closest approach the triangle APC must be right angled with the right angle at apex A of the triangle. This makes the calculation of the length CA simple; CA = 8 km. The other angles in the triangle are $\sin^{-1}(6/10) = 36.86°$ at C and 53.14° at P. The initial positions of

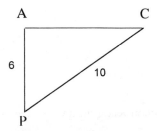

Fig. P.14 The position vector diagram showing the initial position, C, of the cruise ship relative to the port, P. The distance of closest approach, PA, to the port and the course, AC, of the ship are also given.

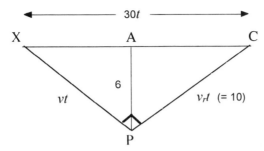

Fig. P.15 The position vector diagram for the interception of the ship by the tug when the tug is travelling with minimum speed for interception to take place.

the ship and the tug are at C and P, respectively, and for the interception to take place the relative velocity of the ship and the tug must be parallel to the line CP. For the interception velocity to be a minimum, the angle between the velocity of the tug and the relative velocity must be a right angle. A position vector diagram for the interception using such a right angled triangle is shown in Fig. P.15. This triangle is not similar to the triangle CPA shown in Fig. P.14; CPA is a component part of the triangle CPX in Fig. P.15. The lengths of the sides of CPX are the velocities multiplied by the time t after the tug leaves port at which the interception takes place. Drawing the triangle in this form enables us to identify $v_r t = CP = 10\,\text{km}$, and we know that the distance AP is $6\,\text{km}$. Using the sine rule first for the triangle XCP and then for the triangle ACP we have, setting v as the tug's velocity,

$$30t = \frac{vt}{\sin 53.14°} = \frac{10}{\sin 53.14°}, \qquad vt = \frac{6}{\sin 53.14°},$$

so that

$$30/v = 10/6 \quad \text{and} \quad v = 18\,\text{kph}.$$

Again, from the triangle XCP, we have

$$\frac{vt}{\sin 53.14°} = \frac{v_r t}{\sin 36.86°},$$

giving

$$v_r = v \tan 53.14° = 24\,\text{kph}$$

and $t = 18/24 = 5/12\,\text{h}$, or $25\,\text{min}$.

If the tug's speed is more than the minimum required for interception to take place, the triangle YCP is no longer right angled and has a form as shown in Fig. P.16. The triangle CPA is unaltered from its previous form, and we can use the cosine rule to express YP in terms of the two other sides of the triangle, as

$$(24t)^2 = (30t)^2 + 100 - 600t \cos 36.86°,$$

which leads to

$$324t^2 - 480t + 100 = 0,$$

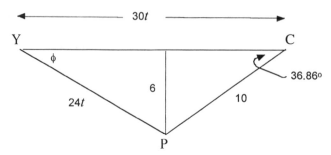

Fig. P.16 The position vector triangle when the tug's speed is more than the minimum required for interception.

a quadratic equation the solutions of which are

$$t = \frac{60 \pm [3600 - 2025]^{1/2}}{81},$$

giving $t = 15\,\text{min}\,3\,\text{s}$ or $1\,\text{h}\,13\,\text{min}\,50\,\text{s}$, so that the time interval in which interception can take place is $58\,\text{min}\,47\,\text{s}$. The angles between the relative velocity vector and the velocity vector of the tug are $85.3°$ and $143.1°$, so that the tug travelling at 24 kph will intercept the cruise ship provided that its course lies between $4.7°$ E of N and $53.1°$ W of N.

16. The space vector diagram requires that the resultant (velocity × time) product, wt, is the third side of a right angled triangle with hypotenuse vt. The minimum value of wt which will close the triangle is (cancelling t)

$$w^2 = v^2 - u^2$$

so that $w > 0$ only if $v > u$.

   The minimum value of the ratio in the sine rule occurs when one of the angles is a right angle. This corresponds with the maximum angle, ϕ, between u and w in velocity space. Then $\sin \phi = 25/40$ and $\phi = 38.68°$. The projectile speed will be 31.2 kph.

17. The velocity–distance relationship is $v^2 = 2ax$, where a is the acceleration. On take-off we have $a = 4$ and $v = 80$, so that $x = 800\,\text{m}$. The stopping distance is 700 m, so that the minimum retardation would be $6400/1400 = 4.57\,\text{m}\,\text{s}^{-2}$.

18. The length of one lap of the circuit is $2[4000 + (\pi \times 500)] = 11\,141.6\,\text{m}$. The previous lap record was obtained at an average speed of $61.11\,\text{m}\,\text{s}^{-1}$ ($= 200\,\text{kph}$) in a time of $11\,141.6/65.11 = 182.3\,\text{s}$. If our car can be driven round the circuit in a time less than 182.3 s, then the lap record will be broken.

   Since the maximum speed on the curves is 180 kph, the acceleration and retardation between 180 and 252 kph must take place in the straight. The time required to travel round the curves is $(1000\pi)/50 = 62.83\,\text{s}$. Entering the straight the driver accelerates to maximum speed in a distance

$$(70^2 - 50^2)/(2 \times 4) = 300\,\text{m}.$$

The same distance will be taken in retardation, so that in one lap 1.2 km is needed to alter speed. This leaves 6.8 km run at $70\,\mathrm{m\,s^{-1}}$, which takes a time of 97.14 s. The time taken in acceleration and deceleration is $4[(70-50)/4] = 20\,\mathrm{s}$. Our car can make a circuit of the track (at maximum performance) in 179.7 s, a new record time. The new record average speed will be 222.87 kph.

19. For the acceleration, $x_1 = 0.5t_1^2$, giving $a_1 = 0.6\,\mathrm{m\,s^{-1}}$, so that $v = 0.6t = 6\,\mathrm{m\,s^{-1}}$. For the retardation $v^2 = 2a_3x_3$, so that $a_3 = 1.2\,\mathrm{m\,s^{-2}}$, $t_3 = 6/1.2 = 5\,\mathrm{s}$, and $t_2 = (400-45)/6 = 59.17\,\mathrm{s}$. The time taken to complete the journey between the two stops is 74.2 s.

20. The restraint exerted on the gun carriage by the restraining harness will be practically zero during the initial part of the gun's recoil; at first the recoiling carriage takes up the slack in the harness. This means that we can deal with the cannon plus cannonball as an isolated system which has no external forces acting on it. The centre of mass of such a system is either static or has a uniform velocity. In this case, before firing, the cannon does not move; after firing the centre of mass of the system is unchanged in position until the restraining force begins to act on the gun carriage. In the absence of external forces the momentum of the components of the system is constant, the momentum with respect to the centre of mass is zero. Taking either of these statements to be applicable to our system we may write

$$MV + mv = 0,$$

where the mass M and the velocity V refer to the cannon and m and v refer to the cannonball, i.e. $10^3 V = -5 \times 400$, so that $V = -2\,\mathrm{m\,s^{-1}}$. The minus sign indicates that the motion of the cannon is in the opposite direction to that of the cannonball.

We can provide a crude justification for our neglect of the restraining force by guessing that the barrel of the cannon is 1 m long. The time interval between the start of firing and the ball's leaving the muzzle will be of the order of $(1/400)\mathrm{s} = 2.5\,\mathrm{ms}$. During that time the cannon will have recoiled by something of the order of 5 mm, not far enough for the restraining force to begin acting.

The average force produced by the harness to bring the cannon to rest can be calculated by equating the kinetic energy of the cannon to the work done by the force, i.e. if the force is F and the stopping distance is 0.4 m,

$$0.4F = \tfrac{1}{2}MV^2 = \tfrac{1}{2}(10^3 \times 4), \qquad \text{so that } F = 5\,\mathrm{kN}.$$

This result could have been derived by calculating the average deceleration as $5\,\mathrm{m\,s^{-2}}$ and multiplying by the cannon mass.

21. The impact of the pile driver on the pile provides an impulse in which momentum is conserved. If the masses of the driver and the pile are M and m respectively, the velocity of the driver on impact is V, and the velocity after impact is v, then

$$MV = (M+m)v,$$

since after the impact the driver and pile move together. The velocity V of the driver is the velocity acquired by a body which has fallen freely under the influence of gravity through a height of 3 m, given by

$$V^2 = 2gh = 2(3 \times 9.8), \qquad \text{so that } V = 7.67 \, \text{m s}^{-1}.$$

The average deceleration of the driver/pile combination is thus $40.85/0.1 = 408.5 \, \text{m s}^{-2}$ and the average force acting is 2.45 MN.

22. A 10 cm rainfall would cover the ground to a depth of 10 cm in the absence of drainage. The mass of water falling on $1 \, \text{m}^2$ of ground in 2 h is $0.1 \times 10^3 \, \text{kg} = 100 \, \text{kg}$, which is the same as $100/7200 = 0.014 \, \text{kg s}^{-1}$. The velocity of the rain when it reaches the ground is given by $v = (2 \times 9.8 \times 300)^{1/2}$ or $v = 76.68 \, \text{m s}^{-1}$. The rate of momentum change for the rain as it hits the ground is

$$0.014 \times 76.68 = 1.07 \, \text{N}.$$

23. The force acting downward on the ball is the weight of the ball, $mg = 4.9 \, \text{N}$. The rate of change of momentum of the water at the underside of the ball must be 4.9 N so that there is no nett force acting on the ball and it remains at rest, supported by the jet. At a height h above the nozzle the velocity, v, of the water jet will be given by $v^2 = 100 - (2 \times 9.8 \times h)$. The rate of change of momentum at the underside of the ball will be $\rho A v^2$, where A is the jet area and ρ is the density of water. Equating this upward acting force with the downward acting force leads to

$$4.9 = 10^3(0.6 \times 10^{-4})(100 - 19.6h), \qquad \text{so that } h = 0.935 \, \text{m}.$$

24. The resistance to the car's motion is provided, effectively, by a column of air travelling at 116 kph relative to the car. The force acting on the car is

$$F = \rho A v^2 = 1.25 \times 1.5(116/3.6)^2 = 1.95 \, \text{kN}.$$

The power produced by the car to overcome this drag force is $P = FV$, where V is the velocity of the car, so that $P = 1.95(100/3.6) \, \text{kW} = 54.1 \, \text{kW}$.

25. The power supplied to the water jet is equal to the rate at which kinetic energy is provided to the jet. In order to calculate this we need to know the mass of water emitted from the jet per second and the velocity of the jet. The rate of volume discharge from the nozzle is $4.8/60 = 0.08 \, \text{m}^3 \, \text{s}^{-1}$, so that the length of the jet emitted per second is $0.08/(6.4 \times 10^{-4}) = 12.5 \, \text{m}$. Since a cross-section of the jet has moved through 12.5 m in 1 s the velocity of the jet is $12.5 \, \text{m s}^{-1}$. The rate of mass discharge is $0.08 \times 10^3 = 80 \, \text{kg s}^{-1}$, so that the rate of supply of kinetic energy is $\frac{1}{2}(80 \times 12.5^2) = 6.25 \, \text{kW}$.

   The force which would be exerted when the jet was incident normally on a wall is $\rho A v^2 = 10^3(6.4 \times 10^{-4}) \times 12.5^2 = 1 \, \text{kN}$.

   It should be noted that the power required to produce the jet isn't obtainable as the product of the force exerted by the jet on the wall and the jet velocity.

26. If the train is to maintain a constant speed, there must be zero nett force acting on it. The force produced by the engine of the train must be 10 kN, and

at 100 kph this is equivalent to a power output of $10^4 \times 100/3.6 = 278\,\text{kW}$. When the train climbs the incline the resistive force is enhanced by the component of the train's weight acting down the slope. Taking the mass of the train as m and the angle defining the slope as α, this has the form $mg \sin \alpha$, with $\tan \alpha = 1/100$. When α is small, as in this case, $\tan \alpha = \alpha = \sin \alpha$, so that the component of the train's weight restraining its motion is $(2 \times 10^5) \times 9.8/100 = 19.6\,\text{kN}$. The total resistive force is $29.6\,\text{kN}$ and the power produced by the engine of the train is $29.6 \times 60/3.6 = 493\,\text{kW}$.

27. After $10\,\text{s}$ the velocity of the car is $18/3.6 = 5\,\text{m s}^{-1}$. The acceleration producing this velocity is $a = 5/10 = 0.5\,\text{m s}^{-2}$, so that the nett force acting on the car of mass m is $0.5m = 500\,\text{N}$. The total force produced by the car engine is more than this by $50\,\text{N}$, so that $F = 550\,\text{N}$. The power output of the car engine is

$$550 \times 5 = 2.75\,\text{kW}.$$

28. The force acting on the bike as it travels at 80 kph is

$$5.6/(80/3.6) = 252\,\text{kN}.$$

The resistive force is $252\,\text{kN}$. When the bike is ascending the slope the component of its weight acting down the slope is $mg \sin \alpha = 150 \times 9.8 \times 0.1 = 147\,\text{N}$. The total force resisting the motion is then $399\,\text{N}$. The velocity of the bike going up the slope will be $5600/399 = 14\,\text{m s}^{-1} = 50.5\,\text{kph}$.

29. The average force acting on the piston of the engine is the average pressure acting on it divided by the area of the piston, $p = F/A$. In this engine, $p = 240\,\text{kN m}^{-2}$ and $A = \pi r^2 = 0.36\pi = 1.13\,\text{m}^2$, so that $F = 271\,\text{kN}$. The power produced by the engine may be taken as the product of the average force acting on the piston and the average velocity of the piston, i.e. $1086\,\text{kW}$.

30. The work done in raising the water is equal to the increase in its total energy, i.e. the sum of its kinetic and potential energies. The potential energy increase of $4.5/60 = 0.075\,\text{m}^3$ of water raised through a height of $15\,\text{m}$ is mgh, where m is the mass of water, ρ is the density of water, and h is the height through which it is raised, so that the rate of increase in potential energy of the water is $(1.125 \times 10^3) \times 9.8 = 11.02\,\text{kJ s}^{-1}$. To find the rate of increase of the kinetic energy of the water, we need to know the velocity with which it leaves the pipe. This is the volume emitted in unit time divided by the cross-sectional area of the pipe, i.e. $0.075/(4 \times 10^{-3}) = 18.75\,\text{m s}^{-1}$. The rate of increase of kinetic energy of the water is then $\frac{1}{2}mv^2 = \frac{1}{2}(0.075 \times 10^3) \times 18.75^2 = 13.18\,\text{kJ s}^{-1}$. The total energy supplied to the water in $1\,\text{s}$ is $24.2\,\text{kJ}$, so that the power required to raise the water is $24.2\,\text{kW}$.

31. The work done in stopping the bullet is the change in its kinetic energy. The kinetic energy is the same as the work done by the uniform forces of resistance to its motion so that $\frac{1}{2}mv^2 = \Sigma F_i x_i$, where m is the mass of the bullet, v is its velocity, and F_i and x_i are the resistive force and the depth of penetration in to the ith material ($i = 1$ or 2). Since we have two sets of

information about the penetration, depending on which side of the armour is struck first by the bullet, we can say, using the subscripts 1 and 2 for steel and plastic respectively, that we can equate the work done in either case, giving

$$0.02F_1 + 0.02F_2 = 0.04F_2 + 0.01F_1 \left(= \tfrac{1}{2}mv^2 \right).$$

The solution of this equation is $F_1 = 2F_2$. Obviously, if the armour is of plastic only $\tfrac{1}{2}mv^2 = 0.06F_2$ and so the penetration depth is 6 cm.

The mass and velocity of the bullet are required only if we need to evaluate the forces and accelerations to which the bullet is subjected. Using the values given above, we find that $F_1 = 133\,\text{kN}$ and $F_2 = 66.6\,\text{kN}$. The assumption that the resistive forces are uniform is naive, since it's hardly likely that the bullet will retain its shape on impact, or that the resistive force is independent of velocity.

32. Initially, when the elevator is static on the ground floor the cage floor supports the weight of the passenger; the force acting on it is $70g = 686\,\text{N}$. The upwards acceleration, a, of the elevator increases the value of g observed by the passenger to $g + a$. If the passenger dropped a dime it would fall under the influence of gravity with an acceleration of g and the floor of the elevator cage would be accelerated towards the falling dime with acceleration a. The two accelerations, although superficially oppositely directed, add together. As the elevator leaves the ground floor the force acting on its floor becomes $70(g + a) = 966\,\text{N}$. When the elevator slows down as it approaches the top floor, its acceleration is directed downwards so that the force acting on the floor of its cage is $70(g - a) = 406\,\text{N}$. In its downward journey the force on the floor is initially 2.9 kN, and as it slows on approaching the ground floor this force becomes 6.9 kN.

An alternative explanation of the way in which a and g are added and subtracted is obtained by considering the contact between the passengers and the floor of the cage during a downward acceleration of the elevator. If the elevator had a downward acceleration greater than g, the passengers would go into free fall inside the elevator cage and would 'float up' towards its roof. Downwards acceleration, although in the same direction as g, reduces the force acting on the cage floor.

33. When the aircraft is flying in a horizontal circle of radius r, the forces acting on it (shown in Fig. P.17) are the normal force N due to a combination of its weight $W (= mg)$, and the centrifugal force (mv^2/r). If there is to be no side slip, the forces parallel to the wings must sum to zero; these forces are the components of mg and mv^2/r parallel to the wings. If the angle of bank is θ, this means that $(mv^2/r)\cos\theta = mg\sin\theta$, or that $\tan\theta = v^2/rg$. The normal force is a combination of the normal components of the weight and centrifugal force, so that

$$N\sin\theta = mv^2/r \qquad \text{and} \qquad N\cos\theta = mg,$$

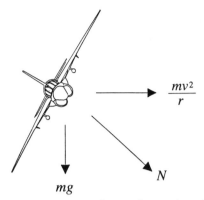

Fig. P.17 The forces acting on the turning aircraft.

which may be expressed as $N^2 = m^2[g^2 + (v^2/r)]$. Since N must be less than $10mg$, the equation to be satisfied for a safe turn is

$$g^2 + (v^2/r) < 10g$$

or $v^2/rg < 99$. This leads to $r = 1780\,\text{m}$. The angle of bank is $84.26°$.

34. The energy loss is simply the reduction in kinetic energy:

$$\tfrac{1}{2}\delta(mv^2) = 0.125(1369 - 1156) = 26.625\,\text{J}.$$

If the ball has travelled in a horizontal straight line (which isn't possible) we could have equated this energy change to the work done by the drag force, F, assuming F to be constant, i.e. $20F = 26.625$, giving $F = 1.33\,\text{N}$. Because the path of the ball from bat to fielder is longer than the straight line joining them, this estimate of the drag force could be said to be the upper limit of the drag force.

35. The change in energy of the frisbee is the change in its total energy, $H = T + V$, i.e. $H = \tfrac{1}{2}mv^2 + mgh$. When it is thrown it has energy H_0, with

$$H_0 = 0.1g + (0.05 \times 12.5^2) = 8.7903\,\text{J}$$

and later

$$H_1 = 0.23g + (0.05 \times 9.6^2) = 6.862\,\text{J},$$

so that the change in energy is $\delta H = H_1 - H_0 = -1.93\,\text{J}$. This loss of energy is a combination of the gain in potential energy, $\delta V = 1.274\,\text{J}$, and the loss in kinetic energy $\delta T = -3.2045\,\text{J}$. The 1.93 J of 'lost' energy have been used in overcoming air drag.

The frisbee is thrown with spin, but we have considered here only the motion of its centre of mass. Since the spin is an intrinsic part of the motion of the frisbee, this is not a particularly good representation of the problem.

36. The volume of the earth, assumed a sphere with radius R, is $\tfrac{4}{3}\pi R^3$. If ρ is the density of coal, the mass of the carbonified earth is $\tfrac{4}{3}\pi R^3\rho$, and it will produce

$35 \times \frac{4}{3}\pi R^3 \rho$ MJ when it is burned. The energy change if the earth were stopped in its orbit would be its loss in kinetic energy:

$$\tfrac{1}{2}MV^2 = \tfrac{2}{3}\pi R^3 \rho (3 \times 10^4) = \tfrac{4}{3}\pi R^3 \rho (450) \text{ MJ}.$$

The ratio of the energy available from stopping the earth to that obtained by burning it is $450/35 = 12.86$. A great deal more energy (nearly 13 times as much) would be obtained if it were stopped in its orbit.

37. It seem plausible to say that the vaulter's centre of mass will be at the height of the bar which he clears. In his run-up his centre of mass is at a height of 1 m relative to the ground, so that he has to raise his centre of mass by 4.5 m if he is to clear the bar. To succeed in his attempt he has to increase his potential energy by $80 \times 4.5 \times 9.8 = 3.53$ kJ. If this potential energy was gained entirely from his kinetic energy, $\frac{1}{2}mv^2 = 40v^2$, the value of his velocity, v, would need to be $v = 3630/40 = 9.39 \text{ m s}^{-1}$. The world record of about 6 m, raising the vaulter's centre of mass by 5 m, would require an approach speed of 9.9 m s^{-1}, which must be very close to the maximum possible while carrying a vaulting pole. It seems likely that both these velocities are overestimates because, as shown in Exercise 4.7, a gymnast in a handstand may well have raised his centre of mass by 90 cm. By adjusting the relative positions of his limbs and torso as he clears the bar, the vaulter could very likely gain an advantage as his centre of mass passes below the bar, perhaps by as much as 50 cm. The 6 m vault would then require a run-up speed of around 9.4 m s^{-1}, which is probably more realistic. Even that may be an overestimate, because the vaulter makes use of the flexibility of the vaulting pole to provide some vertical impulse.

38. The load with a weight, say mg, moves downwards from the position $-l$ relative to the ceiling (defined by the unstretched length, l, of the spring) to the equilibrium position $-(y_0 + l)$. The change in gravitational potential energy for such a change in position is $-mg(y_0 + l) + mgl = -mgy_0$. The extension of the spring to its equilibrium length is defined by $mg = ky_0$, where k is the spring stiffness. The change in gravitational potential energy can be written as ky_0^2. On the other hand, the potential energy of the spring has increased by $\frac{1}{2}ky_0^2$, only half the change in gravitational potential energy. This apparently alarming discrepancy can be explained away quite easily when it's remembered that the gravitational potential energy change, mgy_0, has been derived by assuming that the weight is in free fall, so that the force acting on it is uniformly mg and the work done by that force is mgy_0. In reality, when the spring has been extended by an arbitrary distance y ($<y_0$) the force acting on the load is $mg - ky$, so that the work done in extending the spring to the equilibrium position is

$$\int_0^{y_0} (mg - ky)\mathrm{d}y = mgy_0 - \tfrac{1}{2}ky_0^2 = \tfrac{1}{2}ky_0^2,$$

which is the same as the change in potential energy of the spring. The apparent exception to the First Law of Thermodynamics came about because we glibly assumed that the force acting on the spring was constant. What we took as a constant force problem was actually a variable force problem.

39. To find the force between the two particles we make use of the relationship between force, F, and potential, V, $F = -\nabla V$. Because V is a function of r only, we establish the form of F simply by differentiating V with respect to r:

$$F(r) = -dV/dr = -\{(\lambda/r^2)\exp(-r/r_0) + (\lambda/r_0 r)\exp(-r/r_0)\}$$

$$= [(-\lambda/r_0 r)\exp(-r/r_0)][1 + (r_0/r)].$$

By substituting in the values given for the parameters in this equation, we can find a value for the force for $r = 1.4$ fm. If we use the values as they are given, the result will be expressed in some outlandish unit rather than the standard unit, the newton. We need to convert the parameters to conventional units, writing $\lambda = 70 \times 1.6 \times 10^{-13} \times 10^{-15} = 1.12 \times 10^{-28}$ J m, and $r = r_0 = 1.4 \times 10^{-15}$ m. The force between the two particles is found to be -4.2 kN. For comparison, the Coulomb force between the protons at this distance apart is $e^2/4\pi\varepsilon_0 r^2$, which has the value of 117 N (e is the electronic charge and ε_0 the permittivity of free space). The short-range nature of this inter-nucleon interaction can be seen if we substitute $r = 14$ fm into the equation for F. The force between the nucleons is then -0.028 N, more than 10^5 times smaller than for $r = 1.4$ fm. At the same separation, the Coulomb force has diminished as well, but only to 1.2 N, by a factor of 100.

40. The coefficient of restitution, e, in a collision is the ratio of the relative velocity after the impact to that before the impact. In this case, $v_{r1} = 2.4$ and $v_{r2} = 0.6$, so that $e = 0.25$. The kinetic energy of the two trucks, with masses m_1 and m_2 and velocities v_1 and v_2 respectively, can be written as

$$2T = m_1 v_1^2 + m_2 v_2^2 = \left(\frac{m_1 + m_2}{m_1 + m_2}\right)(m_1 v_1^2 + m_2 v_2^2)$$

i.e.

$$2(m_1 + m_2)T = m_1^2 v_1^2 + m_2^2 v_2^2 + m_1 m_2 v_1^2 + m_2 m_1 v_2^2$$

$$= (m_1 v_1 + m_2 v_2)^2 + m_1 m_2 (v_1 + v_2)^2,$$

so that

$$T = T_c + \frac{m_1 m_2}{2(m_1 + m_2)} v_r^2,$$

where T_c is the energy associated with the motion of the centre of mass of the two trucks. Since there are no external forces acting on the two trucks, T_c is unaltered by the collision. Because $ev_{r1} = -v_{r2}$, the energy loss in the collision is

$$\frac{m_1 m_2}{2(m_1 + m_2)} v_{r1}^2 (1 - e^2) = 9 \text{ kJ}.$$

Since the initial energy of the moving truck was 28.8 kJ, 31.25% of its energy was dissipated; most of it—we could guess—by deformation of the truck bodies.

If we are interested in the actual, rather than the relative, velocities of the trucks after the impact, we invoke the conservation of momentum and the definition of e:

$$m_1v_1 = m_1v_1' + m_2v_2'$$

and

$$v_1 = e(v_2' - v_1')$$

with the result that, after the impact, $v_2' = 2\,\mathrm{m\,s}^{-1}$ and $v_1' = 1.4\,\mathrm{m\,s}^{-1}$.

41. The velocity of the ball can be resolved into a velocity component parallel to the line of the cushion and a component perpendicular to the cushion. The component parallel to the cushion is unaltered by the collision (provided that the cushion is smooth). The perpendicular component is reversed in direction and, unless the collision is elastic, is reduced in magnitude. In this case the perpendicular component of the velocity after impact is $ev\sin(45°) = 1.751\,\mathrm{m\,s}^{-1}$. The parallel component of the velocity, both before and after the impact, is $2.001\,\mathrm{m\,s}^{-1}$. The square of the velocity after impact is $1.751^2 + 2.001^2 = 7.07\,\mathrm{m^2\,s}^{-1}$, so that the loss in kinetic energy given by $2.82^2 - 7.07$ is 0.20 J, nearly 15% of the initial energy. The angle, θ, between the path of the ball and the cushion after the impact is given by $\tan\theta = 1.757/2.0$, i.e. $\theta = 41.19°$. Following the ball through successive collisions with the cushions, first at right angles, and then parallel to, the original cushion would allow a comparison with the case of perfectly elastic collisions in which the track of the ball is part of a rectangular figure.

42. The range, R, of the projectile is related to the angle of projection, θ, by $R = (v^2\sin 2\theta)/g$, where v is the velocity of projection and θ is the angle of projection. Rearranging the equation gives $\sin 2\theta = Rg/v^2 = (10^3 \times 9.8)/125^2 = 0.6272$, which leads to $\theta = 19.4°$ or, remembering that $\sin 2\theta = \sin(\pi - 2\theta)$, $\theta = 70.6°$. The time of flight, T, is obtained from $(v\cos\theta)T = R$, giving $T = 8.48\,\mathrm{s}$ or 24.1 s. If the mortar were fired from the base of the trench, which would be around 2 m below ground level, the larger angle of elevation would be the only possible one unless its crew were prepared to expose themselves to enemy fire. This larger angle would also provide the optimum entry for the mortar bomb into the enemy trench although, if it were spotted, it would give time for the enemy troops to take some evasive action. Another factor which would influence the choice of angle is the allowable error in angle of elevation. If the enemy trench were 2 m wide the bomb would enter the trench provided that the angle of projection didn't deviate from 70.6° by more than ±0.16°, or that it didn't deviate from 19.4° by more than ±0.02°. Obviously, the first of these conditions is easier to satisfy than is the second.

The observation that the two possible angles of projection sum to 90° is a particular example of the relationship between the sum of the angles of projection and the coordinates of a point through which the path of the projectile passes. Let's say that these coordinates are a and b and that the velocity of projection is v. We can write the equation of the trajectory of the projectile in the form

$$y = x \tan \theta - \frac{gx^2}{2v^2} \sec^2 \theta.$$

On substituting $(1 + \tan^2 \theta) = \sec^2 \theta$, $x = a$, $y = b$, and rearranging, we find that the values of θ required if the projectile path is to pass through (a, b) are the solutions of the equation

$$\tan \theta = \frac{v^2}{ga} \pm \left(\frac{v^4}{g^2 a^2} + \frac{2v^2 b}{ga^2} + 1 \right)^{1/2}.$$

If we call these two angles θ_1 and θ_2, then we can write

$$\tan(\theta_1 + \theta_2) = \frac{\tan \theta_1 + \tan \theta_2}{1 + \tan \theta_1 \tan \theta_2} = -a/b.$$

When $b = 0$, for example, for the target trench, $(\theta_1 + \theta_2) = \frac{1}{2}\pi$.

43. When the car is travelling at its maximum speed, all of the power output from its engine is used in overcoming the forces resisting its motion—there is no power available to accelerate it further. We equate power output to the product of resistive force and velocity to obtain

$$22.2 \times 10^3 = 30[200 + k(30^2)],$$

which leads to $k = 0.6\,\mathrm{N\,m^{-2}\,s^2}$. The forces opposing the vehicle's motion when it is travelling up a slope of 1 in 16 are $200 + 0.6v^2$, and the component of its weight parallel to the slope, $mg \sin \alpha$, where α is the angle between the slope and the horizontal. With a gradient of 1 in 16 the angle α is small, so that we can use the approximation $\sin \alpha = \tan \alpha = \frac{1}{16} (= \alpha)$, so that the forces opposing the motion when the velocity is $24\,\mathrm{m\,s^{-1}}$ uphill are $200 + 345.6$ and $900 \times 9.8 \times 0.0625$, which sum to give $1097\,\mathrm{N}$. At a velocity of $24\,\mathrm{m\,s^{-1}}$ this represents a power of $24.32\,\mathrm{kW}$, so that an additional $4.12\,\mathrm{kW}$ would be required to drive the vehicle up the slope at that speed.

44. From Fig. P.18 it can be seen that the normal force on the drum due to the pebble of mass m when it is at the position defined by the angle θ is $mg \cos \theta$. This component of the weight is countered by the centrifugal force mv^2/a, where a is the radius of the drum. The nett normal force acting on the drum is then $mg \cos \theta - mv^2/a$. The pebble will leave the drum when this normal force becomes zero, i.e. when $v^2 = ag \cos \theta$. The energy, H, of the pebble when it is at the position defined by θ is a combination of its potential, V, and kinetic, T, energies. The energy equation may be written

$$2mga = \tfrac{1}{2}mv^2 + mga(1 + \cos \theta)$$

or

$$v^2 = 2ag(1 - \cos \theta).$$

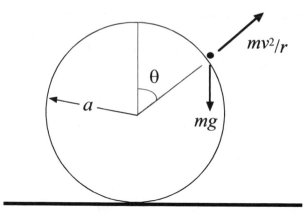

Fig. P.18 The pebble of mass m sliding down the side of the cylindrical drum.

The normal force, then, is zero when $ag \cos \theta = 2ag(1 - \cos \theta)$, which equation is satisfied for $\cos \theta = \frac{2}{3}$, $\theta = 48.19°$. The height above the ground at which this occurs is $a(1 + \cos \theta) = \frac{5}{3}a$, and this is the height at which the pebble leaves the surface of the drum.

When the pebble leaves the drum, its velocity is given by $v^2 = 19.6/3 = 6.53$, i.e. $v = 2.56 \, \text{m s}^{-1}$. Relative to the ground this velocity has a 'perpendicular' component, $v_{y0} = 1.70 \, \text{m s}^{-1}$, and a 'parallel' component $v_x = 1.91 \, \text{m s}^{-1}$. The parallel component of the velocity is a consequence of the normal force at the surface of the drum. Once the pebble leaves the drum, the parallel component of its velocity remains constant. The component of the velocity perpendicular to the ground, v_y, increases at the expense of the pebble's potential energy, i.e.

$$v_y^2 = v_{y0}^2 + 2ga(1 + \cos \theta) = 2.89 + (19.6 \times \tfrac{5}{3}) = 5.96 \, \text{m s}^{-1}.$$

The square of the velocity of the pebble at the ground is formed from the sum of the squares of its components, $v^2 = 35.56 + 3.65$, leading to $v = 6.26 \, \text{m s}^{-1}$. The angle of incidence on the ground is given by $\tan \theta = (v_y/v_x)$, which gives $\theta = 72.28°$ relative to the horizontal. Energy conservation in the gravitational field produces the result that the velocity of the falling pebble is given by $v = (2ga)^{1/2}$, but this is not simply the vertical component of the velocity, as it would be in free fall.

45. The time for which the ball was airborne was $24/3 = 8 \, \text{s}$, and its range of projection was 24 m. We can write the parametric equations for the path of the projectile as

$$x = tV \cos \alpha \quad \text{and} \quad y = tV \sin \alpha - \tfrac{1}{2}gt^2,$$

so that $y = 0$ for $t = 0$ and for $t = T = (2V \sin \alpha)/g$. The value of x corresponding to the time T is the range of projection, $R = (2V^2 \sin \alpha \cos \alpha)/g$. Eliminating V from R and T leads to $gT^2 = 2R \tan \alpha$. Substituting in the figures given, $\tan \alpha = (9.8 \times 9)/48$, i.e. $\alpha = 61.44°$. The velocity of projection was $(Rg/\sin 2\alpha)^{1/2} = 16.74 \, \mathrm{m \, s}^{-1}$. This isn't the complete story. Let's say that the house was symmetrically placed along the path of the ball and that its outside walls were separated by 12 m. The ball would reach a wall position in a time of 0.75 s after projection, and the elevation of the ball would then be 8.27 m. Provided that the walls were of height $h \leqslant 8 \, \mathrm{m}$ (making a small allowance for eaves and gutters), the ball would pass over them. The minimum pitch of the roof would be about 26.5° to the vertical.

46. The angle u defining the slope of the track is $\tan^{-1}(\frac{1}{2}) = 26.565°$. The forces opposing the motion of the car are the component of its weight parallel to the gradient, $mg \sin \theta$, and the frictional force, the product of μ and the component of its weight normal to the slope, $\mu mg \cos \theta$. The tension required to produce an acceleration of $0.5 \, \mathrm{m \, s}^{-2}$ is then the sum of the accelerating force and these resistive forces, i.e.

$$\begin{aligned} \mathbf{T} &= (10^3 \times 0.5) + (10^3 \times 9.8 \times 0.447) + (0.1 \times 10^3 \times 9.8 \times 0.894) \\ &= 5757 \, \mathrm{N}. \end{aligned}$$

While the car is accelerating, an engine the power of which is $v\mathbf{T} = at\mathbf{T}$ is required to provide this force; for $t = 1 \, \mathrm{s}$ this is 2.88 kW and for $t = 2$ it is 5.76 kW. Let's say that the acceleration ceases at $t = 1$ when the velocity of the car is $0.5 \, \mathrm{m \, s}^{-1}$. The power of the motor must be maintained at 2.63 kW to haul the car up the track. By far the larger proportion, 76%, of the work done is done against the gravitational force.

47. The charge Q carried by a capacitor of capacity C is given by $Q = CV$, where V is the voltage (potential difference) across its terminals. When the capacitor is charged by the cell its charge is $1 \, nC \, (= 1 \times 10^{-9}$ coulomb). When the capacitor is connected to the inductance, there is a flow of current and so we have a potential difference across the inductance as well as one across the capacitor. Since there is no source of e.m.f. (i.e. no electric cell) in the circuit, these two potential differences must be equal. The potential difference across and inductance in which the current, I, is changing at a rate of dI/dt is $V_L = -L(dI/dt)$. Current is the rate of flow of charge, $I = dQ/dt$, so that we can write the equation for the potential differences in the LC circuit as

$$L\frac{d^2Q}{dt^2} + \frac{Q}{C} = 0$$

or

$$\frac{d^2Q}{dt^2} = -\frac{1}{LC}Q.$$

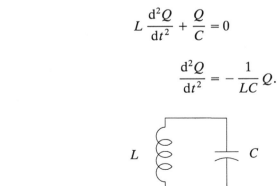

Fig. P.19 The circuit in which a capacitor, C, discharges through an inductance, L.

This equation describing the flow of charge in the circuit is a linear oscillator equation for which the simple harmonic motion has an angular frequency $\omega = (LC)^{-(1/2)}$. The solution of the equation has the form

$$Q = A \sin \omega t + B \cos \omega t,$$

where the constants A and B are determined by the boundary conditions. The condition to be applied here is that $Q = 1\,nC$ when $t = 0$, giving $A = 0$ and $B = 10^{-9}$. Since $\omega = (10^{-9} \times 10^{-5})^{-1/2} = 10^7$, we have $Q = 10^{-9} \cos(10^7 t)$. The current is obtained from Q by differentiating with respect to t to give

$$I = 10^{-9} \times 10^7 \cos(10^7 t) = 10^{-2} \cos(10^7 t).$$

The current is simple harmonic, of amplitude 10 mA, and frequency 1.59 MHz. It's not usual to have a resistance-free circuit, and the equation of motion with a resistance $R\Omega$ in the circuit becomes

$$\frac{d^2 Q}{dt^2} + \frac{R}{L}\frac{dQ}{dt} + \frac{Q}{LC} = 0,$$

which is an equation for a damped harmonic oscillator. The oscillations will be critically damped for $R^2 = 4L/C$, and for values of $R \geqslant (4L/C)^{1/2}$ the discharge will not be oscillatory. In this instance the condition for a non-oscillatory discharge is $R^2 \geqslant 4 \times 10^{-5}/10^{-9}$, i.e. for an oscillatory discharge $R < 100\,\Omega$. For values of $R \geqslant 200\,\Omega$ the charge will be a monotonically decreasing function of time, and so will the current. In this example the exponential decay term $\exp[-(R/2L)t]$ has the form $\exp(-5 \times 10^4 Rt)$. For $R = 20\,\Omega$ the exponent has the value $-10^6 t$, which will have the value -1 when $t = 1\,\mu s$, i.e. by the time $1\,\mu s$ has elapsed the charge and the current will have decayed to $1/e$ of their initial values. In about six cycles of oscillation, the amplitude has reduced nearly to a third of its initial value.

48. At the surface of Bacchus the acceleration due to gravity is $GM/b^2 = \gamma$. At the orbit radius the equality of the gravitational attraction and the centrifugal force gives $ms\omega^2 = GMm/s^2$, so that $\omega = (b^2\gamma/s^3)^{1/2}$. The period of revolution is

$$\mathcal{T} = 2\pi/\omega = 2\pi(s^3/b^2\gamma)^{1/2}.$$

When there are three satellites there will be additional gravitational forces acting on each one due to the presence of the others. Since the moons are arranged at the corners of an equilateral triangle we can say, for example, that the forces acting on Z are

$$ms\omega^2 = \frac{GMm}{s^2} + \frac{2Gm^2 \cos 30°}{(2s \cos 30°)^2}$$

or

$$\omega^2 = \frac{GM}{s^3} + \frac{Gm}{2s^3 \cos 30°},$$

leading to the new period of revolution

$$\tau = \mathcal{T}\left(1 + \frac{m}{M\sqrt{3}}\right)^{1/2}.$$

For a ratio $m/M = 0.01$, this leads to a difference only of about 0.3%.

49. The extension of the spring in the balance by the additional load from its initial equilibrium state to its new equilibrium state is defined by the equality of the extra weight and the change in the tension of the spring. For a spring constant k and an extension x, $kx = mg$. In this case $1.5 \times 9.8 = 0.05k$, i.e. $k = 294\,\mathrm{N\,m^{-1}}$. The equation of motion of the harmonic oscillator is $d^2x/dt^2 = -\omega^2 x$, with $\omega^2 = k/m$. When $m = 6$, as it does when the flour is dropped, $\omega^2 = 49\,\mathrm{s^{-2}}$ and the period of oscillation is $\mathcal{T} = 2\pi/\omega = 0.29\pi$ s or 1.098 s.

When the load is 2.5 cm above its lowest point the extension of the spring from its unstretched state is $x_0 + (5.0 - 2.5)10^{-2} = x_0 + 0.025$ m, where x_0 is the extension of the spring produced by the 6 kg load. The tension in the spring is $k[x_0 + 0.025]$ and, of course, $kx_0 = 6g$, so that the tension, F, in the spring is

$$F = 6g + 0.025k = 66.25\,\mathrm{N}.$$

Because the load is at its lowest position when the oscillation starts, we can write the displacement from the equilibrium position (x_0) as $x = a\cos\omega t$, with $a = 0.05$ and $\omega = 7$. When $x = 0.025$ this defines $\cos\omega t = \frac{1}{2}$, i.e. $\omega t = 60°$. The velocity of the load is $v = -\omega a \sin\omega t$ so that the velocity at $x = 0.025$ m is

$$v = -(7 \times 0.05 \times 0.866) = -0.3\,\mathrm{m\,s^{-1}}.$$

The negative sign indicates that the velocity is directed towards the origin, the initial equilibrium position. The next time that the load is at the same position v will be a positive $0.3\,\mathrm{m\,s^{-1}}$ because $\sin(\pi/3) = -\sin(5\pi/3)$.

One point of interest is that if the 6 kg load had fallen off the balance, leaving the 1.5 kg load on the balance, the amplitude of oscillation would have been so large that the flour bag would have collided with the body of the balance. Alternatively, if an elastic string were used as a support for the load, removing the larger part of the load would lead to the string becoming slack during the early part of the oscillatory motion.

50. Irrespective of the actual form of the pendulum, we make the assumption that we can model it as a simple pendulum for which the period of oscillation is $\mathcal{T} = 2\pi(l/g)^{1/2}$, with l its length and g the acceleration due to gravity. We can say that at the new elevation the sea level value of g has altered by δg so that, if we want to have $\mathcal{T}$ unaltered, we would need to change l by δl so that

$$\frac{l + \delta l}{g + \delta g} = \frac{l}{g},$$

which leads to

$$\delta l/l = \delta g/g.$$

We can write the expression for g as $g = K/r^2$, where K is a constant including the gravitational constant and the mass of the earth. Using this relationship, the change in g with elevation is given by

$$\delta g = K \left(\frac{1}{(R + \delta r)^2} - \frac{1}{R^2} \right)$$

so that

$$\frac{\delta g}{g} = \left(\frac{R}{R + \delta r} \right)^2 - 1 = -\frac{2\,\delta r}{R},$$

where the last step involves our neglecting the order of $(\delta r)^2$. Then

$$(\delta l/l) = -2(\delta r/R) = -(2 \times 1.6)/(6.4 \times 10^3) = -5 \times 10^{-4} = -0.05\%$$

or half a millimetre for each metre of length.

If the length remained unaltered, the fractional change in $\mathscr{T}$ would be $\delta\mathscr{T}/\mathscr{T} = -\frac{1}{2}(\delta g/g) = \delta r/R = 2.5 \times 10^{-4}$. This would represent the clock's losing nearly 22 s per day, or nearly 2 min each week.

This simplistic approach neglects such factors as the non-spherical shape of the earth, and effects such as the gravitational attraction due to variations of the local density of the material of the earth from its average value. The earliest attempts to measure the value of the gravitational constant involved observations of the attraction exerted in its vicinity by a mountain.

51. The parcel of mass m produces a normal force mg on the table. If the coefficient of friction is μ, the force which has to be overcome in order for the parcel to move relative to the table is μmg. While the parcel is moving with the table top, it is performing simple oscillations with the same frequency as does the table top. If we say that the angular frequency of these oscillations is ω, then the equation describing the motion of the parcel is $m(d^2x/dt^2) = -m\omega^2 x$, where x is the displacement of the parcel from its equilibrium position (i.e. its position before the earthquake started). The force acting on the parcel is $-m\omega^2 x$ so that, if the amplitude of oscillation is a, the maximum force acting on the parcel is $m\omega^2 a$. The parcel will move relative to the table unless this maximum force is less than μmg. The condition to be satisfied for the parcel to stay in place $\mu mg > m\omega^2 a$, or $\mu > \omega^2 a/g$. In this case we have a period of oscillation of 1 s, so that $\omega = 2\pi$ and $a = 0.06$ m. The parcel will not move on the table provided that $\mu > (4\pi^2 \times 0.06)/9.8$, i.e. for $\mu > 0.242$. If μ is equal to, or greater than, this value the table top and the parcel move together with no movement relative to one another.

52. The centre of mass of the door is at a distance of h ($= 60$ cm) from the hinge, so that the change in potential energy when the door closes is $mgh = 20 \times 9.8 \times 0.6 = 117.6$ J. Energy must be conserved so that, neglecting dissipative effects at the hinge, the kinetic energy must have increased by 117.6 J. This kinetic energy is $\frac{1}{2}I\omega^2$, where I is the moment of inertia of the door about the hinge and ω is its angular velocity. In a door of height H and mass

m, $I = \frac{1}{3}mh^2$, and in this case $I = 20 \times (1.44/3) = 7.2 \, \text{kg m}^2$. This gives $\omega^2 = 117.6(2/9.6) = 24.5 \, \text{s}^{-2}$, or $\omega = 4.95 \, \text{s}^{-1}$. As the bottom of the door is 1.2 m from the hinge, the velocity of the bottom of the door as it hits the wall is $v = r\omega = 5.94 \, \text{m s}^{-1}$, about 21 kph.

53. The dearth of information about the dimensions of the wheel and axle indicates that the question can be answered only by using the work-energy equality. The cord is 3 m long and a force of 100 N is exerted on it. By the time the cord leaves the axle the force has moved through a distance of 3 m and the work done is 300 J. This must equal the rotational kinetic energy of the system, $\frac{1}{2}I\omega^2$, and with $\omega = 10\pi \, \text{s}^{-1}$ we have $I = 0.61 \, \text{kg m}^2$.

   This must be an imposing demonstration equipment, for if the axle is of small diameter we can say that $I = \frac{1}{2}mr^2$, where m is the disc mass and r is its radius. Here we have $mr^2 = 1.22$, so that if the disc had a mass of 5 kg it would have a diameter of nearly 1 m. For a wheel of mass 50 kg the diameter would be about 31 cm.

54. The forces acting on the rule are shown in Fig. P.20. The torques tending to take the rule to the vertical position are $\frac{1}{2}Mgl \sin \theta$ and $mgl \sin \theta$, so that the nett torque is

$$\Gamma = -\tfrac{1}{2}gl(M + m)\sin \theta.$$

The moment of inertia of the system is that of the rule $\frac{1}{3}Ml^2$ added to that of the mass m, ml^2, i.e.

$$I = \tfrac{1}{3}l^2(M + 3m)$$

so that the equation of motion, $\Gamma = I(d^2\theta/dt^2)$, takes the form

$$\frac{l^2}{3}(M + 3m)\frac{d^2\theta}{dt^2} = -\frac{gl}{3}(M + 2m)\sin \theta.$$

For small values of θ, $\sin \theta = \theta$, and

$$\frac{d^2\theta}{dt^2} = -\frac{g}{l}\frac{3(M + 2m)}{2(M + 2m)}\theta,$$

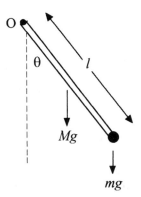

Fig. P.20 The forces acting on a system made from a rigid rod of length l and of mass M, and a mass m, suspended to oscillate about O.

which is the equation for a linear oscillator with angular frequency

$$\omega = \left(\frac{3(M+2m)}{2(M+3m)} \frac{g}{l} \right)^{1/2}.$$

The period of oscillation is $\mathcal{T} = 2\pi/\omega$, so that, in comparison with the simple pendulum of length $1\,\mathrm{m}$ the period of oscillation of which is $\mathcal{T}(1)$,

$$\frac{\mathcal{T}}{\mathcal{T}(1)} = \left(\frac{2(M+3m)}{3(M+2m)} \right)^{1/2} = \left(\frac{(m+\frac{1}{3}M)}{(m+\frac{1}{2}M)} \right)^{1/2}$$

which, with $m = 1$ and $M = 0.15$, gives $\mathcal{T} = 0.988\,\mathcal{T}(1)$, a difference only of 1.2%.

The error introduced into g by the erroneous assumption is 2.4%. Not surprisingly, this error becomes larger as M and m become more nearly equal. For $M/m = 0.2$, the percentage error is 2.4, while for $M/m = 0.6$, it is 7.7%.

55. Provided that the frictional torque acting on the wheel is constant, the angular acceleration α ($= \mathrm{d}\omega/\mathrm{d}t$ with ω the angular velocity of the wheel) will be constant. Since ω in this case is related to time by $\omega = \omega_0 + \alpha t$ and the final angular velocity is zero, we can say that the initial angular velocity $\omega_0 = -\alpha T$, where t is the stopping time. Then $40\pi/6 = -90\alpha$, or $\alpha = -4\pi/54\,\mathrm{s}^{-2}$. The angle through which the wheel has turned while stopping is given by

$$\theta = \omega_0 t + \tfrac{1}{2}\alpha t^2 = 90(40\pi/6) - 100(2\pi/27) = 300\pi.$$

The wheel performs 150 revolutions before stopping.

The equation relating the torque to the acceleration is $\Gamma = I(\mathrm{d}\omega/\mathrm{d}t)$. Using the simple approximation suggested, $I = \tfrac{1}{2}mr^2 = 0.045\,\mathrm{kg\,m}^2$, and

$$\Gamma = 0.045 \times 0.233 = 0.010\,\mathrm{N\,m}.$$

56. The energy, E, supplied in the $10\,\mathrm{s}$ interval is the product of power, P, and time t: $E = Pt = 40\,\mathrm{kJ}$. The energy consumed by a stamping process is then $32\,\mathrm{kJ}$. For most effective operation this energy exchange, taken from the rotational kinetic energy of the flywheel and used in the pressing process, will reduce the rate of revolution of the flywheel from 130 to $100\,\mathrm{rpm}$, i.e. from an angular velocity of $13.6\,\mathrm{s}^{-1}$ to $10.47\,\mathrm{s}^{-1}$. This represents a change in rotational kinetic energy of $\delta T = \tfrac{1}{2}I(\omega_1^2 - \omega_2^2) = 32\,\mathrm{kJ}$, so that $I = (64 \times 10^3)/75.67$, or $I = 846\,\mathrm{kg\,m}^2$. The moment of inertia of the flywheel is $\tfrac{1}{2}Mr^2$, where M is its mass and r its radius. Since $M = 2\,\mathrm{tonnes}$, $846 = 10^3 \times r^2$, or $r = 0.92\,\mathrm{m}$. The flywheel has a diameter of nearly $2\,\mathrm{m}$. Just to obtain a feel for the overall dimensions of the wheel, we can assume that it's made from steel which has a density of between 7 and $8\,\mathrm{tonnes\,m}^{-3}$. The mass of the flywheel is $M = \pi r^2 d\rho$, where d is the width of the wheel and ρ the density of the material from which it is made. This means that for the $2\,\mathrm{tonne}$ wheel of diameter around $2\,\mathrm{m}$, the value of d is about $10\,\mathrm{cm}$. The flywheel has the appearance of a thin disc, but that's only because it's so large.

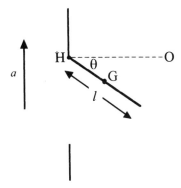

Fig. P.21 A plan view of the railway coach door closing when the train accelerates with an acceleration of a m s^{-2}. The original position of the door is OH and its centre of mass is at G.

57. As the hinge H in Fig. P.21 is accelerated with acceleration a, the door in its initial position OH experiences an acceleration of magnitude a in the opposite direction. The analogy between this situation and the rotation of a rod under the influence of gravity is clear. The torque acting to rotate the door about H is $ma(\frac{1}{2}l \cos \theta)$, where l is the door width and θ is the angle through which it has rotated. The angular motion is described by

$$\tfrac{1}{2}mal \cos \theta = I(d\omega/dt) = I\omega(d\omega/d\theta),$$

so that, on integrating, $\frac{1}{2}mal \sin \theta = \frac{1}{2}I\omega^2 + C$, where C is the constant of integration. Since $\omega = 0$ when $\theta = 0$, $C = 0$. When $\theta = \frac{1}{2}\pi$, when the door will lock itself, if that's possible: $\omega^2 = 4\pi^2/9 = mal/I$. We can write $I = \frac{1}{3}ml^2$, so that the value of a for which the door will close is $a = 4\pi^2l/27 = 0.58\,\mathrm{s}^{-2}$. For the door to shut, it must close towards the rear of the train, which accelerates forwards.

58. From Fig. P.22 we see that the torque acting on the yo-yo to produce the rotation is $Tr = I(d\omega/dt)$, while the force acting to produce the linear acceleration is $mg - T = m(dv/dt) = mr(d\omega/dt)$ since $v = r\omega$. Eliminating $d\omega/dt$ from these equations gives $mg - T = mr^2 T/I$, or

$$mg = T[1 + (mr^2/I)].$$

The mass of the yo-yo, the sum of the masses of the spindle and the capping discs, is 83.4 g. The moment of inertia is the sum of the moments of inertia of the spindle and of the discs and is 5.86×10^{-5} kg m^2. Substituting these figures into the equation gives $T = 0.67\,\mathrm{N}$, to be compared with $mg = 0.81\,\mathrm{N}$. Had we written the moment of inertia as the product of mass and the square of the radius of gyration (K^2) of the yo-yo, $I = mK^2$, the equation for T would be

$$T = mgK^2/(K^2 + r^2).$$

An alternative trick with the yo-yo is to accelerate the end of its string upwards, so that the yo-yo doesn't fall downwards. If it stays at the same

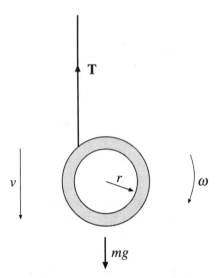

Fig. P.22 The forces (tension in the string and weight of the toy) acting on the yo-yo of mass m and spindle radius r as it rotates with angular velocity ω and falls with velocity v.

position the two forces acting on it in the vertical direction must be equal and opposite: $\mathbf{T_S} = mg$. The rotational motion is described by

$$T_S r = mgr = I(\mathrm{d}\omega/\mathrm{d}t).$$

When the yo-yo doesn't move up or down, the acceleration upwards of the string at the point at which it leaves the spindle must be equal to the rotational acceleration of the yo-yo, since that point doesn't move up or down but rotates with the spindle. Then the acceleration of the end of the string must be $a_s = r(\mathrm{d}\omega/\mathrm{d}t) = mgr^3/I = gr^2/K^2$, which, in this case, has a value of about $2.2\,\mathrm{m\,s^{-2}}$.

59. The kinetic energy of the system is the sum of the kinetic energies of the stone and the rollers. As shown in Fig. P.23, when the block moves in the

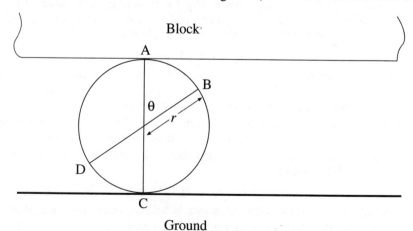

Fig. P.23 The stone/roller system used in building the pyramids.

direction of its velocity, v, by a distance equal to the length of the arc AB ($= r\theta$) the roller has rotated through an angle θ. This rotation has moved the point of contact of the roller with the ground by the length of the arc CD. The roller has moved a distance $r\theta$ relative to the track, while the stone has moved a distance of $2r\theta$ with respect to the track. The linear velocity of the rollers is half that of the stone. The kinetic energy of the system is given by

$$T = \tfrac{1}{2}Mv^2 + 3\left[\tfrac{1}{2}m(v/2)^2\right] + 3\left[\tfrac{1}{2}m(r^2/2)\omega^2\right],$$

where M is the stone mass and m is a roller mass, respectively. If the velocity of the stone is v, then $\omega^2 = (v/2r)^2$ and $T = \tfrac{1}{2}Mv^2 + \tfrac{3}{4}mv^2 = 1.54\,\text{kJ}$.

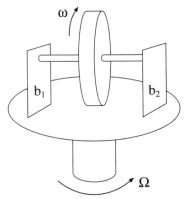

Fig. P.24 The second gyroscope made by the enthusiastic lecturer.

60. The new demonstration gyroscope is illustrated in Fig. P.24. If the radii of the wheel and axle are a_W and a_A, and they have an angular velocity ω, then the angular momentum of the wheel and axle combination is

$$I\omega = \tfrac{1}{2}(Ma_W^2 + ma_A^2)\omega = L = 1.414 \times 10^{-2}\,\text{J s},$$

where M and m are the masses of the wheel and the axle. If the table is rotated about the vertical axis with an angular velocity Ω, then this angular momentum, L, is rotated with angular velocity Ω about the vertical axis. Such a rotation of the angular momentum requires that a torque of magnitude

$$|\Gamma| = |\Omega \times L| = \omega L \sin \theta = \Omega L$$

is produced acting on the gyroscope. The force, F, due to this torque acting vertically on the bearings at the spring balance pans is related to the torque by $\Gamma = d \times F$, where d is the distance from the axle centre to the support. Because the axes of rotation are perpendicular to one another, we can drop the vector notation and say that the forces acting on the spring balances are

$$F_{1,2} = \tfrac{1}{2}[(M + m)g \pm F] = \tfrac{1}{2}[(M + m)g \pm (I\omega\Omega/d)].$$

The force on one of the balances becomes zero for $\Omega = [\{(M + m)gd\}/I\omega]$. The rate of rotation will be specified by Ω, which has the value of $2.287\,\text{s}^{-1}$, about 22 rpm.

Appendix: Calculus

Differentiation

When we can describe a straight line in terms of a coordinate system whose origin is situated on the line we can specify it in terms of the cartesian coordinates, x and y, as

$$y = kx$$

where k is a constant, called the gradient (or slope) of the line. For any increase Δx in x from some arbitrary value, say x_0, the corresponding increase in y from y_0, Δy, is k times Δx. If we chose the point (x_0, y_0) as the origin of the coordinate system, we could express the equation of the line as

$$\Delta y = k\,\Delta x \qquad \text{or as} \quad (\Delta y / \Delta x) = k.$$

In other words, had we started from the point (x_0, y_0) on the line and moved along the line by increasing x by δx (using the lower case δ to indicate a change different from that indicated by Δ), we would have moved to a new position on the line $(x_0 + \delta x, y_0 + \delta y)$ with $\delta y = k\,\delta x$, or $\delta y = \delta x(\Delta y / \Delta x)$. However small we make the increment Δx this last relation remains true. This leads to the idea that there is a limiting value of $(\Delta y / \Delta x)$ as $\Delta x \to 0$,

represented by
$$\underset{\Delta x \to 0}{\text{Limit}} \left(\frac{\Delta y}{\Delta x} \right) = \frac{dy}{dx}$$

which can be said loosely to be the 'value of $(\Delta y / \Delta x)$ at a point'. Then for small increments δx and δy of x and y around the point (x_0, y_0) we can write

$$y_0 + \delta y = x_0 + \delta x \left(\frac{dy}{dx} \right).$$

We call (dy/dx) the first derivative of y with respect to x. $(dy/dx) = k$ is the gradient of the straight line.

We could have said, more formally, that the equation of our straight line expressed y as a function of x, written as $f(x)$, i.e.

$$y = f(x) = kx$$

and that
$$\frac{dy}{dx} = \frac{df}{dx} = k.$$

What if $f(x)$ doesn't have the simple form kx (e.g. $y = x^2$, or x^n, or some sum of powers of x, or some trigonometric function of x)? Can we set up the expressions for the first derivative of $f(x)$ which enable us to write for a microscopic increment, δx, in x

$$y_0 + \delta y = f(x_0) + \delta x \left(\frac{df}{dx} \right)_0$$

where the subscript 0 to the derivative indicates that the first derivative of f has been evaluated for $x = x_0$? The answer, provided that δx is sufficiently small, and that $f(x)$ is a well behaved function, is that we can. It's necessary to evaluate the first derivative for $x = x_0$ because, except in the case when we are dealing with straight lines, the first derivative, (df/dx) is a function which involves x so that it has different values for different values of x.

Example S1. What is the first derivative with respect to x of the function

$$f(x) = y = x^2?$$

We are interested here in the way in which $f(x)$ changes when x increases infinitesimally by δx. The new value of x is $(x + \delta x)$ so that

$$f(x) + \delta f(x) = (x + \delta x)^2 = x^2 + 2x\delta x + (\delta x)^2.$$

Provided that $\delta x \ll 1$ (and it is chosen as very small indeed, infinitesimally small), $(\delta x)^2 \ll \delta x$, so that we can drop the term $(\delta x)^2$ without affecting the result appreciably. This 'neglect of the second order of small quantities' way leads to

$$f(x) + \delta f(x) = x^2 + 2x\delta x$$

or

$$\delta f(x) = 2x\delta x$$

since $f(x) = x^2$. On dividing through by δx and taking the limiting value as $\delta x \to 0$ we obtain

$$\frac{df(x)}{dx} = 2x = \frac{dy}{dx}.$$

The first derivative of x^2 is $2x$. By using a similar argument involving the expansion of $[1 + (\delta x/x)]^n$ as a power series in $(\delta x/x)$ (by means of the binomial theorem) it is straightforward to show that the first derivative of x^n is nx^{n-1}.

Example S2. If $y = \sin x$ what is (dy/dx)?

This time $y + \delta y = f(x) + \delta f(x) = \sin(x + \delta x) = \sin x \cos \delta x + \cos x \sin \delta x$. Since when δx is small we can substitute $\sin \delta x = \delta x$, and $\cos \delta x = 1$, we are left with $y + \delta y = \sin x + \delta x \cos x$ which leads to $(dy/dx) = \cos x$. The way in which we have defined the first derivative indicates that, when the function $f(x)$ is other than linear, (dy/dx) is the slope of the tangent to the curve at the value of x under consideration. The tangent to a curve is horizontal when $(dy/dx) = 0$, and this case is of particular interest because it can represent a maximum or a minimum of the curve. Let's look at the case of a curve $y = f(x)$ which shows a minimum at a particular value of x, say at $x = a$. For $x < a$ the slope of the curve is negative, for $x > a$ it is positive, whilst for $x = a$ it is zero. The slope of a line representing the gradient of the tangent to the curve $f(x)$ is positive (i.e. a graph of

(df/dx) plotted as a function of x is a line with a positive slope). The derivative for such a line, obtained by differentiating (df/dx) with respect to x will be positive. This derivative of the derivative of $f(x)$ is denoted by $(d^2f(x)/dx^2)$ and is called the second derivative of $f(x)$. It is equally easy to show that $(d^2f(x)/dx^2)$ is negative at the position of a maximum in $f(x)$.

Example S3. What is the first derivative with respect to x of $y = f(x) = x \sin x$.

By substituting $(x + \delta x)$ in place of x we obtain

$$f(x + \delta x) = (x + \delta x)\sin(x + \delta x) = (x + \delta x)(\sin x \cos \delta x + \cos x \sin \delta x)$$

$$= x \sin x + x \delta x \cos x + \delta x \sin x + \cos x (\delta x)^2$$

so that, neglecting the term involving $(\delta x)^2$, and identifying $x \sin x$ with $f(x)$, we have

$$(df(x)/dx) = \sin x + x \cos x.$$

If we set $x = V(x)$ and $\sin x = U(x)$ so that $f(x) = V(x)U(x)$ this equation could be written as

$$\frac{df(x)}{dx} = \frac{d[V(x)U(x)]}{dx} = U(x)\frac{dV(x)}{dx} + V(x)\frac{dU(x)}{dx}$$

or, more briefly,

$$\frac{df}{dx} = U\frac{dV}{dx} + V\frac{dU}{dx}$$

which is a general result for a function which is a product of two other functions. (Example S3 could be solved by setting $V(x) = x$ and $U(x) = \sin x$.)

Example S4. x and y can be written in parametric form as $x = t^2$ and $y = t^4$. What is the form of (dy/dx)?

The simplicity of the parametric forms of x and y makes it obvious that we can write $y = x^2$ so that $(dy/dx) = 2x$. On the other hand, we have $(dy/dt) = 4t^3$ and $(dx/dt) = 2t$. If we divide the first of these by the second we obtain

$$\frac{dy}{dt}\frac{dt}{dx} = (4t^3/2t) = 2t^2 = 2x,$$

the result we obtained above. This suggests that we can relate the derivative of y with respect to x to the derivatives of y and of x with respect to the parameter t by using the 'chain rule of differentiation'

$$\frac{dy}{dx} = \frac{dy}{dt}\frac{dt}{dx}.$$

(This rule is demonstrated rather than proved in this example because its proof is long-winded.)

The operation of differentiation is associative; the derivative of the sum of two functions is the sum of the individual derivatives of the two functions.

Integration

Integration is a process which is the inverse of differentiation. Suppose that we know the derivative (df/dx) of a function $f(x)$; is it possible for us to establish the form of $f(x)$? If we divide the region of x which is of interest into small increments δx, and for each increment multiply the derivative of $f(x)$ by δx we obtain a set of δf's, one for each δx. When we add up all the values of δf in the region of interest, we obtain the form of $f(x)$. Because x is continuously variable this summation is not simply a summation of discrete numbers; it is no longer called summation but goes by the name of integration. Symbolically,

$$f(x) = \lim_{\delta x \to 0} \sum \frac{\delta f}{\delta x} \delta x \quad \text{denoted by} \quad \int \frac{df}{dx} dx = \int df + b$$

where b is a constant. If we are to solve the integral we need to know the boundary conditions which will enable us to evaluate b. Because the product $(\delta f/\delta x)\delta x$ can be visualized as the area of an infinitesimally thin rectangle with sides δx and $(\delta f/\delta x)$, the process of integration, the 'summation' of all the products $(df/dx)dx$, represents the evaluation of the area underneath the curve of (df/dx) plotted as a function of x.

Example S4.   Evaluate the integral $\int x^n\, dx$.

Since integration is the inverse of differentiation, we can solve this problem provided that we can identify the function whose derivative is x^n. The derivative of x^{n+1} is $(n+1)x^n$ so that the derivative of $[x^{n+1}/(n+1)]$ is x^n. We then write

$$\int x^n\, dx = [x^{n+1}/(n+1)] + b.$$

The constant b is necessary because when we visualize the process of integration as the evaluation of the area under a curve we cannot say what is its value unless we specify the values of x at either end of the region of x which is of interest. When we know what these values are, say p and q, we say that p and q are the limits between which the integral is evaluated and write the integral as

$$\int_p^q x^n\, dx = \left. \frac{x^{n+1}}{n+1} \right|_p^q = \frac{1}{n+1}[q^{n+1} - p^{n+1}].$$

When the integral has limits in this way it is known as a *definite* integral, as opposed to the *indefinite* integral which has no limits specified. The constant b has cancelled out on the left hand side of the equation; citing p and q provides us with the boundary conditions necessary to evaluate the integral. Obviously the integral of a function between the limits p and q has the same magnitude, but the opposite sign, to the same integral between the limits q and p. Exchanging the limits on the integral sign changes the sign, but not the magnitude, of the result of the integration. The sum of two integrals of the same function taken between the limits p and q and between the limits q and r is the same as the integral of the function taken between the limits p and r.

Index